PRAISE FOR

What a Plant Knows

"Plants may be brainless, eyeless and devoid of senses as we know them, but they have a rudimentary 'awareness,' says biologist Daniel Chamovitz. In this beautiful reframing of the botanical, he reveals the extent and kind of that awareness through a bumper crop of research." —*Nature*

"A handy guide to our own senses as well as those of plants." —Frank Graham Jr., *Audubon*

"By comparing human senses to the abilities of plants to adapt to their surroundings, [Chamovitz] provides a fascinating and logical explanation of how plants survive despite the inability to move from one site to another. Backed by new research on plant biology, this is an intriguing look at a plant's consciousness." —*Kirkus Reviews*

"Verdict: Plant-astic." —*Herald Sun* (Australia)

"Chamovitz's book is pop science at its best, full of vivid examples of barely imaginable ways of living." —Peter Marren, *BBC Wildlife*

"In a lively and delightful discourse that aligns botany with human biology, [Chamovitz] articulates his findings about plants and the senses in accessible, often whimsical observations that make complex science not only comprehensible but fun to ponder." —Carol Haggas, *Booklist*

"Thick with eccentric plant experiments and astonishing plant science." —James McConnachie, *The Sunday Times* (London)

"[Chamovitz] gently hints that we should have a greater appreciation of plants' complexity and perceptiveness . . . If plants can see, smell, feel, know where they are, and remember, then perhaps they do possess some kind of intelligence. Maybe that is worth reflecting on the next time you casually stroll past a plant."
—Chelsie Eller, *Science*

"Like us, a plant that aspires to win the rat race must exploit its environment. Even a daffodil can detect when you're standing in its light, and a rhododendron knows when you're savaging its neighbor with the pruning shears. With deftness and clarity, Chamovitz introduces plants' equivalent of our senses, plus floral forms of memory and orientation. When you realize how much plants know, you may think twice before you bite them."
—Hannah Holmes, author of *Quirk* and *Suburban Safari*

"Just as his groundbreaking research uncovered connections between the plant and animal kingdoms, Chamovitz's insights in *What a Plant Knows* transcend the world of plants. This entertaining and educational book is filled with wondrous examples that underscore how the legacy of shared genomes enables plants and animals to respond to their environments. You'll see plants in a new light after reading *What a Plant Knows*."
—Gloria M. Coruzzi, Carroll and Milton Petrie Professor, Center for Genomics and Systems Biology, New York University

"If you've ever marveled at how and why plants make the choices they do, *What a Plant Knows* holds your answer. Chamovitz is a master at translating the science of botany into the language of the layman."
—Michael Malice, author, subject of Harvey Pekar's *Ego & Hubris*, and succulent enthusiast

"Chamovitz walks the *Homo sapiens* reader right into the shoes—or I should say roots—of the plant world. After reading this book you will never again walk innocently past a plant or reach insensitively for a leaf. You will marvel at and be haunted by a plant's sensory attributes and the shared genes between the plant and animal kingdoms."

—Elisabeth Tova Bailey, author of
The Sound of a Wild Snail Eating

"A fascinating book that explores accessibly the evidence that plants share more properties with animals than most people appreciate. It may come as a relief to vegetarians to learn that plants do not feel pain or suffer, in the human sense, when harvested. Nevertheless, after reading *What a Plant Knows*, we wanted to apologize to our daffodils for the times when our shadows have shielded them from the sun."

—Mary and John Gribbin, authors of *Flower Hunters*

"*What a Plant Knows* is lively, eloquent, scientifically accurate, and easy to read. I commend this engaging text to all who wonder about life on earth and seek a compelling introduction to the lives of plants as revealed through centuries of careful scientific experimentation."

—Professor Stephen D. Hopper, Director,
Royal Botanic Gardens, Kew

DANIEL CHAMOVITZ
What a Plant Knows

Daniel Chamovitz, Ph.D., is the director of the Manna Center for Plant Biosciences at Tel Aviv University. He has served as a visiting scientist at Yale University and at the Fred Hutchinson Cancer Research Center, and has lectured at botanical gardens around the world. His work has been featured in *The Boston Globe*, *The Wall Street Journal*, and *The Daily Beast*, and on NPR, the BBC, and more. Chamovitz lives with his wife and three children in Hod HaSharon, Israel. You can visit his website at www.danielchamovitz.com.

What a Plant Knows

What a Plant Knows

A FIELD GUIDE TO THE SENSES

Daniel Chamovitz

SCIENTIFIC AMERICAN / FARRAR, STRAUS AND GIROUX

NEW YORK

Scientific American / Farrar, Straus and Giroux
18 West 18th Street, New York 10011

A portion of chapter 2 originally appeared, in slightly different form, in *Scientific American*.

The Library of Congress has cataloged the hardcover edition as follows:
Chamovitz, Daniel, 1963–
    What a plant knows : a field guide to the senses / Daniel Chamovitz. — 1st ed.
      p. cm.
    Includes bibliographical references and index.
    ISBN 978-0-374-28873-0 (alk. paper)
    1. Plants.   2. Plant physiology.   I. Title.

  QK50 .C45 2012
  571.2—dc23

                                                  2011040179

Paperback ISBN: 978-0-374-53388-5

Designed by Jonathan D. Lippincott

www.fsgbooks.com · books.scientificamerican.com
www.twitter.com/fsgbooks · www.facebook.com/fsgbooks

10 9

For Shira, Eytan, Noam, and Shani

Contents

What a Plant Knows

Prologue

My interest in the parallels between plant and human senses got its start when I was a young postdoctoral fellow at Yale University in the 1990s. I was interested in studying a biological process specific to plants and not connected to human biology (probably as a response to the six other doctors in my family, all of whom are physicians). Hence I was drawn to the question of how plants use light to regulate their development. In my research I discovered a unique group of genes necessary for a plant to determine if it's in the light or in the dark. Much to my surprise and against all of my plans, I later discovered that this same group of genes is also part of the human DNA. This led to the obvious question as to what these seemingly "plant-specific" genes do in people. Many years later and after much research we now know that these genes not only are conserved between plants and animals but also regulate (among other developmental processes) responses to light in both!

This led me to realize that the genetic difference between plants and animals is not as significant as I had once believed. I began to question the parallels between plant and human biology even as my own research evolved from studying plant responses to light to leukemia in fruit flies. What I discovered was that while there's no plant that knows how to say "Feed me, Seymour!" there are many plants that "know" quite a bit.

Indeed, we tend not to pay much attention to the immensely sophisticated sensory machinery in the flowers and trees that can be found right in our own backyards. While most animals can choose their environments, seek shelter in a storm, search for food and a mate, or migrate with the changing seasons, plants must be able to withstand and adapt to constantly changing weather, encroaching neighbors, and invading pests, without being able to move to a better environment. Because of this, plants have evolved complex sensory and regulatory systems that allow them to modulate their growth in response to ever-changing conditions. An elm tree has to know if its neighbor is shading it from the sun so that it can find its own way to grow toward the light that's available. A head of lettuce has to know if there are ravenous aphids about to eat it up so that it can protect itself by making poisonous chemicals to kill the pests. A Douglas fir tree has to know if whipping winds are shaking its branches so that it can grow a stronger trunk. Cherry trees have to know when to flower.

On a genetic level, plants are more complex than many animals, and some of the most important discoveries in all of biology came from research carried out on plants. Robert Hooke first discovered cells in 1665 while studying cork in the early microscope he built. In the nineteenth century Gregor Mendel worked out the principles of modern genetics using pea plants, and in the mid-twentieth century Barbara McClintock used Indian corn to show that genes can *transpose*, or jump. We now know that these "jumping genes" are a characteristic of all DNA and are intimately connected to cancer in humans. And while we recognize that Darwin was a founding father of modern evolutionary theory, some of his most important findings were in *plant* biology specifically, and we'll see quite a few of these in the pages of this book.

Clearly, my use of the word "know" is unorthodox. Plants don't have a central nervous system; a plant doesn't have a brain that coordinates information for its entire body. Yet different parts of a plant are intimately connected, and information regarding light, chemicals in the air, and temperature is constantly exchanged between roots and leaves, flowers and stems, to yield a plant that is optimized for its environment. We can't equate human behavior to the ways in which plants function in their worlds, but I ask that you humor me while I use terminology throughout the book that is usually reserved for human experience. When I explore what a plant *sees* or what it *smells*, I am not claiming that plants have eyes or noses (or a brain that colors all sensory input with emotion). But I believe this terminology will help challenge us to think in new ways about sight, smell, what a plant is, and ultimately what we are.

My book is not *The Secret Life of Plants*; if you're looking for an argument that plants are just like us, you won't find it here. As the renowned plant physiologist Arthur Galston pointed out back in 1974 during the height of interest in this extremely popular but scientifically anemic book, we must be wary of "bizarre claims presented without adequate supporting evidence." Worse than leading the unwary reader astray, *The Secret Life of Plants* led to scientific fallout that stymied important research on plant behavior as scientists became wary of any studies that hinted at parallels between animal senses and plant senses.

In the more than three decades since *The Secret Life of Plants* caused a great media stir, the depth at which scientists understand plant biology has increased immensely. In *What a Plant Knows*, I will explore the latest research in plant biology and argue that plants do indeed have senses. By no means is this book an exhaustive and complete review of what modern science has to say about plant senses; that would necessitate a textbook inaccessible to all but the most dedicated readers. Instead, in each chapter I highlight a human sense and compare what the sense is for people and what it is for plants. I describe how the sensory information is perceived, how it is processed, and the ecological implications of the sense for a plant. And in each chapter I'll present both a historical perspective and a modern look at the topic. I've chosen to cover sight, feeling, hearing, proprioception, and memory, and while I'll devote a chapter to smell, I'm not focusing on taste here—the two senses are intimately connected.

We are utterly dependent on plants. We wake up in houses made of wood from the forests of Maine, pour a cup of coffee brewed from coffee beans grown in Brazil, throw on a T-shirt made of Egyptian cotton, print out a report on paper, and drive our kids to school in cars with tires made of rubber that was grown in Africa and fueled by gasoline derived from cycads that died millions of years ago. Chemicals extracted from plants reduce fever (think of aspirin) and treat cancer (Taxol). Wheat sparked the end of one age and the dawn of another, and the humble potato led to mass migrations. And plants continue to inspire and amaze us: the mighty sequoias are the largest singular, independent organisms on earth, algae are some of the smallest, and roses definitely make anyone smile.

Knowing what plants do for us, why not take a moment to find out more about what scientists have found out about them? Let's embark on our journey to explore the science behind the inner lives of plants. We'll start by uncovering what plants really see while they're hanging out in the backyard.

ONE

What a Plant Sees

She turns, always, towards the sun, though her roots
hold her fast, and, altered, loves unaltered.

—Ovid, *Metamorphoses*

Think about this: plants see you.

In fact, plants monitor their visible environment all the time. Plants see if you come near them; they know when you stand over them. They even know if you're wearing a blue or a red shirt. They know if you've painted your house or if you've moved their pots from one side of the living room to the other.

Of course plants don't "see" in pictures as you or I do. Plants can't discern between a slightly balding middle-aged man with glasses and a smiling little girl with brown curls. But they do see light in many ways and colors that we can only imagine. Plants see the same ultraviolet light that gives us sunburns and infrared light that heats us up. Plants can tell when there's very little light, like from a candle, or when it's the middle of the day, or when the sun is about to set into the horizon. Plants know if the light is coming from the left, the right, or from above. They know if another plant has grown over them, blocking their light. And they know how long the lights have been on.

So, can this be considered "plant vision"? Let's first examine what vision is for us. Imagine a person born blind, living in total darkness. Now imagine this person being given the ability to discriminate between light and shadow. This person could differentiate between night and day, inside and outside. These new senses would definitely be considered rudimentary sight and would enable new levels of function. Now imagine this person being able to discern color. She can see blue above and green below. Of course this would be a welcome improvement over darkness or being able to discern only white or gray. I think we can all agree that this fundamental change—from total blindness to seeing color—is definitely "vision" for this person.

Merriam-Webster's defines "sight" as "the physical sense by which light stimuli received by the eye are interpreted by the brain and constructed into a representation of the position, shape, brightness, and usually color of objects in space." We see light in what we define as the "visual spectra." Light is a common, understandable synonym for the electromagnetic waves in the visible spectrum. This means that light has properties shared with all other types of electrical signals, such as micro- and radio waves. Radio waves for AM radio are very long, almost half a mile in length. That's why radio antennas are many stories tall. In contrast, X-ray waves are very, very short, one trillion times shorter than radio waves, which is why they pass so easily through our bodies.

Light waves are somewhere in the middle, between 0.0000004 and 0.0000007 meter long. Blue light is the shortest, while red light is the longest, with green, yellow, and orange in the middle. (That's why the color pattern of rainbows is always oriented in the same direction—from the colors with short waves, like blue, to the colors with long waves, like red.) These are the electromagnetic waves we "see" because our eyes have special proteins called photoreceptors that know how to receive this energy, to absorb it, the same way that an antenna absorbs radio waves.

The retina, the layer at the back of our eyeballs, is covered with rows and rows of these receptors, sort of like the rows and rows of LEDs in flat-screen televisions or sensors in digital cameras. Each point on the retina has photoreceptors called rods, which are sensitive to all light, and photoreceptors called cones, which respond to different colors of light. Each cone or rod responds to the light focused on it. The human retina contains about 125 million rods and 6 million cones, all in an area about the size of a passport photo. That's equivalent to a digital camera with a resolution of 130 megapixels. This huge number of receptors in such a small area gives us our high visual resolution. For comparison, the highest-resolution outdoor LED displays contain only about 10,000 LEDs per square meter, and common digital cameras have a resolution of only about 8 megapixels.

Rods are more sensitive to light and enable us to see at night and under low-light conditions but not in color. Cones allow us to see different colors in bright light since cones come in three flavors—red, green, and blue. The major difference between these different photoreceptors is the specific chemical they contain. These chemicals, called rhodopsin in rods and photopsins in cones, have a specific structure that enables them to absorb light of different wavelengths. Blue light is absorbed by rhodopsin and the blue photopsin; red light by rhodopsin and the red photopsin. Purple light is absorbed by rhodopsin, blue photopsin, and red photopsin, but *not* green photopsin, and so on. Once the rod or cone absorbs the light, it sends a signal to the brain that processes all of the signals from the millions of photoreceptors into a single coherent picture.

Blindness results from defects at many stages: from light perception by the retina due to a physical problem in its structure; from the inability to sense the light (because of problems in the rhodopsin and photopsins, for example); or in the ability to transfer the information to the brain. People who are color-blind for red, for example, don't have any red cones. Thus the red signals are not absorbed and passed on to the brain. Human sight involves cells that absorb the light, and the brain then processes this information, which we in turn respond to. So what happens in plants?

Darwin the Botanist

It's not widely known that for the twenty years following his publication of the landmark *On the Origin of Species*, Charles Darwin conducted a series of experiments that still influence research in plants to this day.

Darwin was fascinated by the effects of light on plant growth, as was his son Francis. In his final book, *The Power of Movement in Plants*, Darwin wrote: "There are extremely few [plants], of which some part . . . does not bend toward lateral light." Or in less verbose modern English: almost all plants bend toward light. We see that happen all the time in houseplants that bow and bend toward rays of sunshine coming in from the window. This behavior is called phototropism. In 1864 a contemporary of Darwin's, Julius von Sachs, discovered that blue light is the primary color that induces phototropism in plants, while plants are generally blind to other colors that have little effect on their bending toward light. But no one knew at that time how or which part of a plant sees the light coming from a particular direction.

In a very simple experiment, Darwin and his son showed that this bending was due not to photosynthesis, the process whereby plants turn light into energy, but rather to some inherent sensitivity to move toward light. For their experiment, the two Darwins grew a pot of canary grass in a totally dark room for several days. Then they lit a very small gas lamp twelve feet from the pot and kept it so dim that they "could not see the seedlings themselves, nor see a pencil line on paper." But after only three hours, the plants had obviously curved toward the dim light. The curving always occurred at the same part of the young plant, an inch or so below the tip.

**Canary grass
(Phalaris canariensis)**

This led them to question which part of the plant saw the light. The Darwins carried out what has become a classic experiment in botany. They hypothesized that the "eyes" of the plant were found at the seedling tip and *not* at the part of the seedling that bends. They checked phototropism in five different seedlings, illustrated by the following diagram:

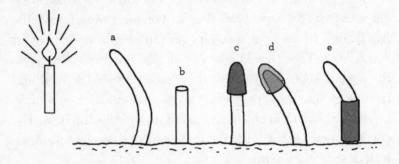

Summary of Darwin's experiments on phototropism

a. The first seedling was untreated and shows that the conditions of the experiment are conducive to phototropism.
b. The second had its tip pruned off.
c. The third had its tip covered with a lightproof cap.
d. The fourth had its tip covered with a clear glass cap.
e. The fifth had its middle section covered by a lightproof tube.

They carried out the experiment on these seedlings in the same conditions as their initial experiment, and of course the untreated seedling bent toward the light. Similarly, the seedling with the lightproof tube around its middle (see *e* above) bent toward the light. If they removed the *tip* of a seedling, however, or covered it with a lightproof cap, it went blind and couldn't bend toward the light. Then they witnessed the behavior of the plant in scenario four (*d*): this seedling continued to bend toward the light even though it had a cap on its tip. The difference here was that the cap was clear. The Darwins realized that the glass still allowed the light to shine onto the tip of the plant. In one simple experiment, published in 1880, the two Darwins proved that phototropism is the result of light hitting the tip of a plant's shoot, which sees the light and transfers this information to the plant's midsection to tell it to bend in that direction. The Darwins had successfully demonstrated rudimentary sight in plants.

Maryland Mammoth: The Tobacco That Just Kept Growing

Several decades later, a new tobacco strain cropped up in the valleys of southern Maryland and reignited interest in the ways that plants see the world. These valleys have been home to some of America's greatest tobacco farms since the first settlers arrived at the end of the seventeenth century. Tobacco farmers, learning from the Native tribes such as the Susquehannock, who had grown tobacco for centuries, would plant their crop in the spring and harvest it in late summer. Some of the plants weren't harvested for their leaves and made flowers that provided the seed for the next year's crop. In 1906, farmers began to notice a new strain of tobacco that never seemed to stop growing. It could reach fifteen feet in height, produce almost a hundred leaves, and would only stop growing when the frosts set in. On the surface, such a robust, ever-growing plant would seem a boon to tobacco farmers. But as is so often the case, this new strain, aptly named Maryland Mammoth, was like the two-faced Roman god Janus. On the one hand, it never stopped growing; on the other, it rarely flowered, meaning farmers couldn't harvest seed for the next year's crop.

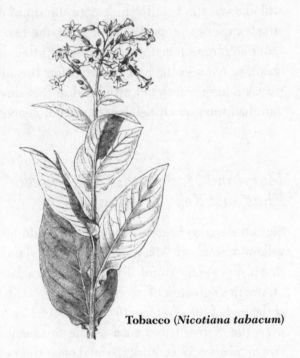

Tobacco (*Nicotiana tabacum*)

In 1918, Wightman W. Garner and Harry A. Allard, two scientists at the U.S. Department of Agriculture, set out to determine why Maryland Mammoth didn't know when to stop making leaves and start making flowers and seeds instead. They planted the Maryland Mammoth in pots and left one group outside in the fields. The other group was put in the field during the day but moved to a dark shed every afternoon. Simply limiting the amount of light the plants saw was enough to cause Maryland Mammoth to stop growing and start flowering. In other words, if Maryland Mammoth was exposed to the long days of summer, it would keep growing leaves. But if it experienced artificially shorter days, then it would flower.

This phenomenon, called photoperiodism, gave us the first strong evidence that plants measure how much light they take in. Other experiments over the years have revealed that many plants, just like the Mammoth, flower only if the day is short; they are referred to as "short-day" plants. Such short-day plants include chrysanthemums and soybeans. Some plants need a long day to flower; irises and barley are considered "long-day" plants. This discovery meant that farmers could now manipulate flowering to fit their schedules by controlling the light that a plant sees. It's not surprising that farmers in Florida soon figured out that they could grow Maryland Mammoth for many months (without the effects of frost encountered in Maryland) and that the plants would eventually flower in the fields in midwinter when the days were shortest.

What a Difference a (Short) Day Makes

The concept of photoperiodism sparked a rush of activity among scientists who were brimming with follow-up questions: Do plants measure the length of the day or the night? And what color of light are plants seeing?

Around the time of World War II, scientists discovered that they could manipulate when plants flowered simply by quickly turning the lights on and off in the middle of the night. They could take a short-day plant like the soybean and keep it from making flowers in short days if they turned on the lights for only a few minutes in the middle of the night. On the other hand, the scientists could cause a long-day plant like the iris to make flowers even in the middle of the winter (during short days, when it shouldn't normally flower), if in the middle of the night they turned on the lights for just a few moments. These experiments proved that what a plant measures is not the length of the *day* but the length of the continuous period of darkness.

Using this technique, flower farmers can keep chrysanthe-mums from flowering until just before Mother's Day, which is the optimal time to have them burst onto the spring flower scene. Chrysanthemum farmers have a problem since Mother's Day comes in the spring but the flowers normally blossom in the fall as the days get shorter. Fortunately, chrysanthemums grown in greenhouses can be kept from flowering by turning on the lights for a few minutes at night throughout the fall and winter. Then . . . *boom* . . . two weeks before Mother's Day, the farmers stop turning on the lights at night, and all the plants start to flower at once, ready for harvest and shipping.

These scientists were curious about the color of light that the plants saw. What they discovered was surprising: the plants, and it didn't matter which ones were tested, only responded to a flash of red during the night. Blue or green flashes during the night wouldn't influence when the plant flowered, but only a few sec-onds of *red* would. Plants were differentiating between colors: they were using blue light to know which direction to bend in and red light to measure the length of the night.

Then, in the early 1950s, Harry Borthwick and his colleagues in the USDA lab where Maryland Mammoth was first studied made the amazing discovery that far-red light—light that has wavelengths that are a bit longer than bright red and is most often seen, just barely, at dusk—could cancel the effect of the red light on plants. Let me spell this out more clearly: if you take

irises, which normally don't flower in long nights, and give them
a shot of red light in the middle of the night, they'll make flowers
as bright and as beautiful as any iris in a nature preserve. But if
you shine far-red light on them right after the pulse of red, it's as
if they never saw the red light to begin with. They won't flower.
If you then shine red light on them after the far-red, they will.
Hit them again with far-red light, and they won't. And so on. We're
also not talking about lots of light; a few seconds of either color is
enough. It's like a light-activated switch: The red light turns on
flowering; the far-red light turns it off. If you flip the switch
back and forth fast enough, nothing happens. On a more philosophical level, we can say that the plant remembers the last color
it saw.

By the time John F. Kennedy was elected president, Warren L. Butler and colleagues had demonstrated that a single photoreceptor in plants was responsible for both the red and the
far-red effects. They called this receptor "phytochrome," meaning "plant color." In its simplest model, phytochrome is the light-
activated switch. Red light activates phytochrome, turning it into
a form primed to receive far-red light. Far-red light inactivates
phytochrome, turning it into a form primed to receive red light.
Ecologically, this makes a lot of sense. In nature, the last light
any plant sees at the end of the day is far-red, and this signifies to
the plant that it should "turn off." In the morning, it sees red
light and it wakes up. In this way a plant measures how long ago
it last saw red light and adjusts its growth accordingly. Exactly
which part of the plant sees the red and far-red light to regulate
flowering?

We know from Darwin's studies of phototropism that the
"eye" of a plant is in its tip while the response to the light occurs

in the stem. So we might conclude, then, that the "eye" for photoperiodism is also in the tip of the plant. Surprisingly, this isn't the case. If in the middle of the night you shine a beam of light on different parts of the plant, you discover that it's sufficient to illuminate any single leaf in order to regulate flowering in the *entire* plant. On the other hand, if all the leaves are pruned, leaving only the stem and the apex, the plant is blind to any flashes of light, even if the entire plant is illuminated. If the phytochrome in a single leaf sees red light in the middle of the night, it's as if the entire plant were illuminated. Phytochrome in the leaves receives the light cues and initiates a mobile signal that propagates throughout the plant and induces flowering.

Blind Plants in the Age of Genetics

We have four different types of photoreceptors in our eyes: rhodopsin for light and shadows, and three photopsins for red, blue, and green. We also have a fifth light receptor called cryptochrome that regulates our internal clocks. So far we've seen that plants also have multiple photoreceptors: they see directional blue light, which means they must have at last one blue-light photoreceptor, now known as phototropin, and they see red and far-red light for flowering, which points to at least one phytochrome photoreceptor. But in order to determine just how many photoreceptors plants possess, scientists had to wait for the era of molecular genetics, which began several decades after the discovery of phytochrome.

The approach spearheaded in the early 1980s by Maarten Koornneef at Wageningen University in Holland, and repeated and refined in numerous labs, used genetics to understand plant sight. Koornneef asked a simple question: What would a "blind" plant look like? Plants grown in darkness or dim light are taller than those grown in bright light. If you ever took care of bean sprouts for a sixth-grade science experiment, you'd know that the plants in the hall closet grew up tall, spindly, and yellow, but the ones out on the playground were short, vigorous, and green. This makes sense because plants normally elongate in darkness, when they're trying to get out of the soil into the light or when they're in the shade and need to make their way to the unobstructed light. If Koornneef could find a blind mutant plant, perhaps it would be tall in bright light as well. If he could identify and grow blind mutant plants, he would be able to use genetics to figure out what was wrong with them.

He carried out his experiments on *Arabidopsis thaliana*, a small laboratory plant similar to wild mustard. He treated a batch of arabidopsis seeds with chemicals known to induce mutations in DNA (and also cause cancer in laboratory rats) and then grew the seedlings under various colors of light and looked for mutant seedlings that were taller than the others. He found many of them. Some of the mutant plants grew taller under blue light, but were of normal height when grown under red light. Some were taller under red light but normal under blue. Some were taller under UV light but normal under all other kinds, and some were taller under red *and* blue lights. A few were taller only under dim light, while others were taller only under bright-light conditions.

Many of these mutants that were blind to specific colors of light were defective in the particular photoreceptors that absorb the light. A plant that had no phytochrome grew in red light as if it were in the dark. Surprisingly, a few of the photoreceptors came in pairs, with one being specific for dim light and the other

Arabidopsis
(*Arabidopsis thaliana*)

specific for bright light. To make a long and complex story short, we now know that arabidopsis has at least *eleven* different photoreceptors: some tell a plant when to germinate, some tell it when to bend to the light, some tell it when to flower, and some let it know when it's nighttime. Some let the plant know that there's a lot of light hitting it, some let it know that the light is dim, and some help it keep time.*

So plant vision is much more complex than human sight at the level of perception. Indeed, light for a plant is much more than a signal; plants need light to eat. Plants use light to turn

---

*More specifically, arabidopsis has at least eleven different photoreceptors that fall into five distinct classes (phototropins, phytochromes, and cryptochromes, plus two additional classes). Other plants also contain these five classes, but may have more or fewer members of each class.

water and carbon dioxide into sugars that in turn provide food for all animals. But plants are sessile, unmoving organisms as well. They are literally rooted in one place, unable to migrate in search of food. To compensate for this sessile life, plants must have the ability to find their food—to seek out and capture light. That means they need to know where the light is, and rather than moving toward the food, as an animal would, a plant grows toward its food.

A plant needs to know if another plant has grown above it, filtering out the light for photosynthesis. If a plant senses that it is in the shade, it will start growing faster to get out. And plants need to survive, which means they need to know when to "hatch" out of their seeds and when to reproduce. Many types of plants start growing in the spring, just as many mammals give birth then. How do plants know when the spring has started? Phytochrome tells them that the days are getting progressively longer. Plants also flower and set seed in the fall before the snow comes. How do they know it's autumn? Phytochrome tells them that the nights are getting longer.

What Plants and Humans See

Plants must be aware of the dynamic visual environment around them in order to survive. They need to know the direction, amount, duration, and color of light to do so. Plants undoubtedly detect visible (and invisible) electromagnetic waves. While we can detect electromagnetic waves in a relatively tight spectrum, plants detect ones that are both shorter and longer than those we can detect. Although plants see a much larger spectrum than we do, they don't see in pictures. Plants don't have a nervous system that translates light signals into pictures. Instead, they translate light signals into different cues for growth. Plants don't have eyes, just as we don't have leaves.*

But we can both detect light.

Sight is the ability not only to *detect* electromagnetic waves but also the ability to *respond* to these waves. The rods and cones in our retinas detect the light signal and transfer this information to the brain, and we respond to the information. Plants are also able to translate the visual signal into a physiologically recognizable instruction. It wasn't enough that Darwin's plants saw the light in their tips; they had to absorb this light and then somehow translate it into an instruction that told the plant to bend. They needed to *respond* to the light. The complex signals arising from multiple photoreceptors allow a plant to optimally modulate its growth in changing environments, just as our four photoreceptors allow our brains to make pictures that enable us to interpret and respond to our changing environments.

To put things in a broader perspective: plant phytochrome and human red photopsin are not the same photoreceptor; while they both absorb red light, they are different proteins with different chemistries. What we see is mediated through photoreceptors found only in other animals. What a daffodil sees is mediated through photoreceptors found only in plants. But the plant and human photoreceptors are similar in that they all consist of a protein connected to a chemical dye that absorbs the light; these are the physical limitations required for a photoreceptor to work.

But there are exceptions to every rule, and despite billions of years of independent evolution plant and animal visual systems

---

*Green algae, the most primitive forms of plants, have an organelle termed the eyespot that allows the algal cells to sense changes in light direction and intensity. These eyespots have been considered the simplest forms of eyes in nature.

do have some things in common. Both animals and plants contain blue-light receptors called cryptochromes.° Cryptochrome has no effect on phototropism in plants, but it plays several other roles in regulating plant growth, one of which is its control over a plant's internal clock. Plants, like animals, have an internal clock called the "circadian clock" that is in tune with normal day-night cycles. In our case, this internal clock regulates all parts of our life, from when we're hungry, when we have to go to the bathroom, when we're tired, and when we feel energetic. These daily changes in our body's behavior are called circadian rhythms, because they continue on a roughly twenty-four-hour cycle even if we keep ourselves in a closed room that never gets sunlight. Flying halfway around the world puts our circadian clock out of sync with the day-night signals, a phenomenon we call jet lag. The circadian clock can be reset by light, but this takes a few days. This is also why spending time outside in the light helps us recover from jet lag faster than spending time in a dark hotel room.

Cryptochrome is the blue-light receptor primarily responsible for the resetting of our circadian clocks by light. Cryptochrome absorbs blue light and then signals the cell that it's daytime. Plants also have internal circadian clocks that regulate

---

°The name "cryptochrome" is actually the result of a joke made by Jonathan Gressel of the Weizmann Institute. Gressel had been studying blue-light responses in a group of organisms that include lichens, mosses, ferns, and algae, which are also called cryptogamic plants (this name will be significant, as we'll see in a second). But like all other researchers studying the effects of blue light on different creatures, he didn't know what the receptor was for blue light. Despite various attempts over decades, no one had succeeded in isolating this receptor; it had a cryptic nature. An unabashed punster, Gressel suggested calling the unidentified photoreceptor "cryptochrome." To the chagrin of many of his colleagues, his joke has become enshrined in scientific nomenclature, even though cryptochrome is no longer cryptic, as it was finally isolated in 1993.

many plant processes, including leaf movements and photosynthesis. If we artificially change a plant's day-night cycle, it also goes through jet lag (but doesn't get grumpy), and it takes a few days for it to readjust. For example, if a plant's leaves normally close in the late afternoon and open in the morning, reversing its light-dark cycle will initially lead to its leaves opening in the dark (at the time that used to be day) and closing in the light (at the time that used to be night). This opening and closing of leaves will readjust to the new light-dark patterns within a few days.

The plant cryptochrome, just like the cryptochrome in fruit flies and mice, has a major role in coordinating external light signals with the internal clock. At this basic level of blue-light control of circadian rhythms, plants and humans "see" in essentially the same way. From an evolutionary perspective, this amazing form of conservation of cryptochrome function is actually not so surprising. Circadian clocks developed early in evolution in single-celled organisms, before the animal and the plant kingdoms split off. These original clocks probably functioned to protect the cells from damage induced by high UV radiation. In this early clock, an ancestral cryptochrome monitored the light environment and relegated cell division to the night. Relatively simple clocks are even found today in most single-celled organisms, including bacteria and fungi. The evolution of light perception continued from this one common photoreceptor in all organisms and diverged into the two distinct visual systems that distinguish plants from animals. What may be more surprising, though, is that plants also smell . . .

TWO

What a Plant Smells

Stones have been known to move and trees to speak.
—Shakespeare, *Macbeth*

Plants smell. Plants obviously emit odors that animals and human beings are attracted to, but they also sense their *own* odors and those of neighboring plants. Plants know when their fruit is ripe, when their neighbor has been cut by a gardener's shears, or when their neighbor is being eaten by a ravenous bug; they smell it. Some plants can even differentiate the smell of a tomato from the smell of wheat. Unlike the large spectrum of visual input that a plant experiences, a plant's range for smell is limited, but it is highly sensitive and communicates a great deal of information to the living organism.

If you look up the word "smell" in a standard dictionary today, you'll see that it is defined as the ability "to perceive odor or scent through stimuli affecting the olfactory nerves." Olfactory nerves can easily be understood as the nerves that connect the smell receptors in the nose to the brain. In olfaction, the stimuli are small molecules dissolved in the air. Human olfaction involves the cells in our nose that receive airborne chemicals, and it involves our brain, which processes this information so that we can respond to various smells. If you open a bottle of Chanel N°5 on one side of a room, for example, you smell it on the other because certain chemicals evaporate from the perfume and disperse across the room. The molecules are present in very dilute quantities, but our noses are filled with thousands of receptors that react specifically with different chemicals. It only takes one molecule to connect with a receptor to sense the new smell.

Our body's mechanism for the perception of smells is different from the mechanism involved in the perception of light. As we saw in the previous chapter, we only need four classes of photoreceptors that differentiate between red, green, blue, and white to see the colors of a complete palette. When it comes to olfaction, however, we have hundreds of different receptor types, each specifically designated for a unique volatile chemical.

The way that a smell receptor in the nose binds to a chemical is similar to the concept of a lock-and-key system. Each chemical has its own particular shape that fits into a specific protein receptor, just as each key has its own particular structure that fits into a specific lock. Only a unique chemical can bind to a unique receptor, and once this happens, it initiates a cascade of signals that ends in a nerve firing in the brain to let us know that the receptor has been stimulated. We interpret this as a particular smell. Scientists have recorded hundreds of individual aroma chemicals such as menthol (the major component of aroma in peppermint) and putrescine (responsible for the foul-smelling aroma that emanates from dead flesh). But any particular aroma we smell is usually the result of a mix of several chemicals. For example, while about half of the peppermint odor is due to menthol, the rest is a combination of more than thirty other chemicals. That's why we can describe the bouquet of an excellent spaghetti sauce, or of a deep red wine, or of a newborn baby in so many different ways.

So what happens in a plant? Our dictionary's definition of "smell" excludes plants from the discussion. They are removed from our traditional understandings of the olfactory world because they do not have a nervous system, and olfaction for a plant is obviously a nose-less process. But let's say we tweak this definition to "the ability to perceive odor or scent through stimuli." Plants are indeed more than remedial smellers. What odors does a plant perceive, and how do smells influence a plant's behavior?

Unexplained Phenomena

My grandmother didn't study plant biology or agriculture. She didn't even finish high school. But she knew that she could get a hard avocado to soften by putting it in a brown paper bag with a ripe banana. She learned this magic from her mother, who learned it from her mother, and so on. In fact, this practice goes back to antiquity, and ancient cultures had diverse methods for getting fruit to ripen. The ancient Egyptians slashed open a few figs in order to get an entire bunch to ripen, and in ancient China people would burn ritual incense in a storage room of pears to get the fruit to ripen.

In the early twentieth century, farmers in Florida would ripen citrus in sheds heated by kerosene. These farmers were sure that the heat induced the ripening, and of course their conclusion sounds logical. You can imagine their dismay, then, when they plugged in some electric heaters near the citrus and found that the fruit didn't cooperate at all. So if it wasn't the heat, could the ripening magic be coming from the kerosene?

It turned out that it was. In 1924, Frank E. Denny, a scientist from the U.S. Department of Agriculture in Los Angeles, demonstrated that kerosene smoke contains minute amounts of a molecule called ethylene and that treating any fruit with pure ethylene gas is enough to induce ripening. The lemons he studied were so sensitive to ethylene that they could respond to a tiny amount in the air, at a ratio of 1 to 100 million. Similarly, it turns out that the smoke from the Chinese incense also contained ethylene. So a simple scientific model would posit that the fruit "smells" minuscule amounts of ethylene in the smoke and translates this smell into rapid ripening. We smell the smoke from a neighbor's barbecue, and we salivate; a plant detects some ethylene in the air, and it softens up.

But this explanation doesn't answer two important questions: First, why do plants respond to the ethylene in smoke anyway? And second, what about my grandma putting two fruits together in a bag and the Egyptians slashing their figs? Experiments carried out by Richard Gane in Cambridge in the 1930s point to some answers. Gane analyzed the air immediately surrounding ripening apples and showed that it contained ethylene. A year after his pioneering work, a group at the Boyce Thompson Institute at Cornell University proposed that ethylene is the universal plant hormone responsible for fruit ripening. In fact, numerous subsequent studies have revealed that all fruits, including figs, emit this organic compound. So it's not just smoke that contains ethylene; normal fruit emits this gas as well. When the Egyptians slashed their figs, they allowed the ethylene gas to easily escape. When we put a ripe banana in a bag with a hard pear, for example, the banana gives off ethylene, which is "smelled" by the pear, and the pear quickly ripens. The two fruits are communicating their physical states to each other.

Ethylene signaling between fruits didn't evolve so that we can have perfectly ripe pears whenever we crave them, of course. Instead, this hormone evolved as a regulator of plant responses to environmental stresses such as drought and wounding and is produced naturally throughout the life cycle of all plants (including little mosses). But ethylene is particularly important for plant aging as it is the major regulator of leaf senescence (the aging process that produces autumn foliage) and is produced in *copious* amounts in ripening fruit. The ethylene produced in ripening apples ensures not only that the entire fruit ripens uniformly but that neighboring apples will also ripen, which will give off even more ethylene, leading to an ethylene-induced ripening cascade of McIntoshes. From an ecological perspective, this has an advantage in ensuring seed dispersal as well. Animals are attracted to "ready-to-eat" fruits like peaches and berries. A full display of soft fruits brought on by the ethylene-induced wave guarantees an easily identifiable market for animals, which then disperse the seeds as they go about their daily business.

Finding Food

Cuscuta pentagona is not your normal plant. It's a spindly orange vine that can grow up to three feet high, produces tiny white flowers of five petals, and is found all over North America. What's unique about *Cuscuta* is that it has no leaves and it isn't green because it lacks chlorophyll, the pigment that absorbs solar energy, which allows plants to turn light into sugars and oxygen through photosynthesis. *Cuscuta* obviously can't carry out photosynthesis, as most plants do, so it doesn't make its own food using light. With all of this in mind, we'd assume *Cuscuta* would starve, but instead it thrives. *Cuscuta* lives in another way: it gets its food from its neighbors. It is a parasitic plant. In order to live, *Cuscuta* attaches itself to a host plant and sucks out the nutrients provided by the host by burrowing an appendage into the plant's vascular system. Not surprisingly, *Cuscuta*, commonly known as the dodder plant, is an agricultural nuisance and is even classified as a "noxious weed" by the U.S. Department of Agriculture. But what makes *Cuscuta* truly fascinating is that it has culinary preferences: it chooses which neighbors to attack.

Before we get to the reasons why *Cuscuta* has very specific and refined culinary tastes, let's see how it starts its parasitical

Five-angled dodder (*Cuscuta pentagona*)

life. *Cuscuta* seeds germinate like any other plant seeds. Placed on soil, the seed breaks open, the new shoot grows into the air, and the new root burrows into the dirt. But a young dodder left on its own will die if it doesn't quickly find a host to live off of. As a dodder seedling grows, it moves its shoot tip in small circles, probing the surroundings the way we do with our hands when we're blindfolded or searching for the kitchen light in the middle of the night. While these movements seem random at first, if the dodder is next to another plant (say, a tomato), it's quickly obvious that the *Cuscuta* is bending and growing and rotating in the direction of the tomato plant that will provide it with food. The dodder bends and grows and rotates until finally it finds a tomato leaf. But rather than touch the leaf, the dodder sinks down and keeps moving until it finds the *stem* of the tomato plant. In a final act of victory, it twirls itself around the stem, sends microprojections into the tomato's phloem (the vessels that carry the plant's sugary sap), and starts siphoning off sugars so that it can keep growing and eventually flower. Oh, and the tomato plant starts to wilt as the dodder thrives.

Dr. Consuelo De Moraes even documented this behavior on film.° She is an entomologist at Penn State University whose main interest is understanding volatile chemical signaling between insects and plants, and between plants themselves. One of her projects centered on figuring out how *Cuscuta* locates its prey. She demonstrated that the dodder vines never grow toward empty pots or pots with fake plants in them but faithfully grow toward tomato plants no matter where she put them—in the light, in the shade, wherever. De Moraes hypothesized that the

---

°To fully appreciate this, you really have to see it with your own eyes: www.youtube.com /watch?v=NDMXvwa0D9E.

dodder actually *smelled* the tomato. To check her hypothesis, she and her students put the dodder in a pot in a closed box and put the tomato in a second closed box. The two boxes were connected by a tube that entered the dodder's box on one side, which allowed the free flow of air between the boxes. The isolated dodder always grew toward the tube, suggesting that the tomato plant was giving off an odor that wafted through the tube into the dodder's box and that the dodder liked it.

If the *Cuscuta* was really going after the smell of the tomato, then perhaps De Moraes could just make a tomato perfume and see if the dodder would go for that. She created an *eau de tomato* stem extract that she placed on cotton swabs and then put the swabs on sticks in pots next to the *Cuscuta*. As a control, she put some of the solvents that she used to make the tomato perfume on other swabs of cotton and put these on sticks next to the *Cuscuta* as well. As predicted, she tricked the dodder into growing toward the cotton giving off the tomato smell, thinking it was going to find food, but not to the cotton with the solvents.

Clearly, the dodder can smell a plant to find food. But as I explained earlier, this noxious weed has its preferences. Given a choice between a tomato and some wheat, the dodder will choose the tomato. If you grow your dodder in a spot that is equidistant between two pots—one containing wheat, the other containing tomato—the dodder will go for the tomato. Even at the level of fragrance, and not the whole plant, dodder prefers *eau de tomato* to *eau de wheat*.

At the basic chemical level, *eau de tomato* and *eau de wheat* are rather similar. Both contain beta-myrcene, a volatile compound (one of the hundreds of unique chemical smells known) that on its own can induce *Cuscuta* to grow toward it. So why the preference? One clear hypothesis is the complexity of the bouquet. In addition to beta-myrcene, the tomato gives off two other volatile chemicals that the dodder is attracted to, making for an overall irresistible dodder-attracting fragrance. Wheat, however, only contains one dodder-enticing odor, the beta-myrcene, and not the other two found in the tomato. What's more, wheat not only makes fewer attractants but also makes (Z)-3-Hexenyl acetate, which repels the dodder more than the beta-myrcene attracts it. In fact, the *Cuscuta* grows *away* from (Z)-3-Hexenyl acetate, finding the wheat simply repulsive.

(L)eavesdropping

In 1983, two teams of scientists published astonishing findings related to plant communication that revolutionized our understanding of everything from the willow tree to the lima bean. The scientists claimed that trees warn each other of imminent leaf-eating-insect attack. The results were relatively straightforward; the implications astounding. News of their work soon spread to popular culture, with the idea of "talking trees" found in the pages not only of *Science* but of mainstream newspapers around the world.

David Rhoades and Gordon Orians, two scientists from the University of Washington, noticed that caterpillars were less likely to forage on leaves from willow trees if these trees neighbored other willows already infested with tent caterpillars. The healthy trees neighboring the infested trees were resistant to the caterpillars because, as Rhoades discovered, the leaves of the resistant trees, but not of susceptible ones isolated from the infested trees, contained phenolic and tannic chemicals that made them unpalatable to the insects. As the scientists could detect no physical connections between the damaged trees and their healthy neighbors—they did not share common roots, and their branches did not touch—Rhoades proposed that the attacked trees must be sending an airborne pheromonal message to the healthy trees. In other words, the infested trees signaled to the neighboring healthy trees, "Beware! Defend yourselves!"

White willow (*Salix alba*)

Just three months later, the Dartmouth researchers Ian Baldwin and Jack Schultz published a seminal paper that supported the Rhoades report. Baldwin and Schultz had been in contact with Rhoades and designed their experiment to be carried out under very controlled conditions, rather than monitoring trees grown open in nature, as Rhoades and Orians had done. They studied poplar and sugar maple seedlings (about a foot tall) grown in airtight Plexiglas cages. They used two cages for their experiment. The first contained two populations of trees: fifteen trees that had two leaves torn in half, and fifteen trees that were not damaged. The second cage contained the control trees, which of course were not damaged. Two days later, the remaining leaves on the damaged trees contained increased levels of a number of chemicals, including toxic phenolic and tannic compounds that are known to inhibit the growth of caterpillars. The trees in the

White poplar (*Populus alba*)

control cage didn't show increases in any of these compounds. The important result here was that the leaves of the *intact* trees in the same cage as the damaged ones also showed large increases in phenolic and tannic compounds. Baldwin and Schultz proposed that the damaged leaves, whether by tearing as in their experiments or by insect feeding as in Rhoades's observations of the willow trees, emitted a gaseous signal that enabled the damaged trees to communicate with the undamaged ones, which resulted in the latter defending themselves against imminent insect attack.

These early reports of plant signaling were often dismissed by other individuals in the scientific community as lacking the correct controls or as having correct results but exaggerated implications. At the same time, the popular press embraced the idea of "talking trees" and anthropomorphized the conclusions of the researchers. Whether it was the *Los Angeles Times* or *The Windsor Star* in Canada or *The Age* in Australia, news outlets went berserk over the idea and carried stories with titles like "Scientists Turn New Leaf, Find Trees Can Talk" and "Shhh. Little Plants Have Big Ears," and the front page of the *Sarasota Herald-Tribune* bore the headline "Trees Talk, Respond to Each Other, Scientists Believe." *The New York Times* even titled its main editorial on June 7, 1983, "When Trees Talk," in which the writer speculated about "talking trees whose bark is worse than their blight." All this public attention didn't help convince scientists to adopt the ideas of chemical communication being posited by Baldwin and his colleagues. But over the past decade, the phenomenon of plant communication through smell has been shown again and again for a large number of plants, including barley, sagebrush, and alder, and Baldwin, a young chemist just barely out of college at the time of the original publication, has gone on to a prominent scientific career.[*]

While the phenomenon of plants being influenced by their neighbors through airborne chemical signals is now an accepted scientific paradigm, the question remains: Are plants truly communicating with each other (in other words, purposely warning each other of approaching danger), or are the healthy ones just eavesdropping on a soliloquy by the infested plants, which do not intend to be heard? When a plant releases a smell in the air, is it a form of talking, or is it, so to say, just passing gas? While the idea of a plant calling out for help and warning its neighbors has allegorical and anthropomorphic beauty, does it really reflect the original intent of the signal?

Martin Heil and his team at the Center for Research and Advanced Studies in Irapuato, Mexico, have been studying wild lima beans (*Phaseolus lunatus*) for the past several years to further explore this question. Heil knew that when a lima bean plant is eaten by beetles, it responds in two ways. The leaves that are being eaten by the insects release a mixture of volatile chemicals into the air, and the flowers (though not directly attacked by the beetles) produce a nectar that attracts beetle-eating arthropods.[†] Early in his career at the turn of the millennium, Heil had worked at the Max Planck Institute for Chemical Ecology in Germany, the same institute where Baldwin was (and still is) a director, and like Baldwin before him Heil wondered why it was that lima beans emitted these chemicals.

[*]Baldwin now directs the Department of Molecular Ecology at the Max Planck Institute for Chemical Ecology in Jena, Germany.
[†]Many insect-eating arthropods have coevolved with plants and recognize the volatile signals emitted by herbivore-attacked plants and use this signal as a food-finding cue.

Heil and his colleagues placed lima bean plants that had been attacked by beetles next to plants that had been isolated from the beetles and monitored the air around different leaves. They chose a total of four leaves from three different plants: from a single plant that had been attacked by beetles they chose two leaves, one leaf that had been eaten and another that was not; a leaf from a neighboring but healthy "uninfested" plant; and a leaf from a plant that had been kept isolated from any contact with beetles or infested plants. They identified the volatile chemical in the air surrounding each leaf using an advanced technique known as gas chromatography–mass spectrometry (often featured on the show *CSI* and employed by perfume companies when they are developing a new fragrance).

Wild lima bean (*Phaseolus lunatus*)

Heil found that the air emitted from the foraged and the healthy leaves on the same plant contained essentially identical volatile chemicals, while the air around the control leaf was clear of these gases. In addition, the air around the healthy leaves from the lima beans that neighbored beetle-infested plants also contained the same volatiles as those detected from the foraged plants. The healthy plants were also less likely to be eaten by beetles.

In this set of experiments, Heil confirmed the earlier studies by showing that the proximity of the undamaged leaves to the attacked leaves provided them with a defensive advantage against the insects. But Heil was not convinced that damaged plants "talk" to other plants to warn them against impending attack. Rather, he proposed that the neighboring plant must be practicing a form of olfactory eavesdropping on an internal signal actually intended for other leaves on the same plant.

Heil modified his experimental setup in a simple, albeit ingenious, way to test his hypothesis. He kept the two plants next to each other but enclosed the attacked leaves in plastic bags for twenty-four hours. When he checked the same four types of leaves as in the first experiment, the results were different. While the attacked leaf continued to emit the same chemical as it did before, the other leaves on the same vine and neighboring vines now resembled the control plant; the air around the leaves was clear.

Heil and his team opened the bag around the attacked leaf, and with the help of a small ventilator that's usually used on tiny microchips to help cool computers, they blew the air in one of two directions: either toward the neighboring leaves farther up the vine or away from the vine and into the open. They checked the gases coming out of the leaves higher up the stem and measured how much nectar they produced. The leaves blown with

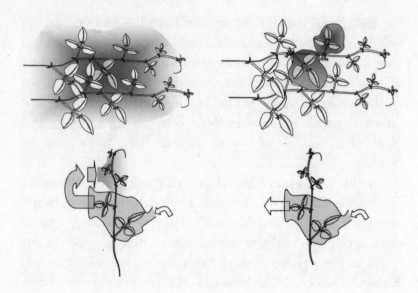

An illustration of Heil's experiments. On the top two panels Heil let beetles attack the gray leaves and then checked the air around other leaves on both the same vine and the neighboring vine. On the top left we see that the air around all the leaves of both vines contained the same chemicals, while on the top right, when Heil isolated the attacked leaves in plastic bags, the air around them was different from the air around all the other leaves on both vines. On the bottom we see his second experiment. Heil blew air from attacked leaves either onto other leaves on the same vine (on the left) or away from the other leaves (the bottom right panel).

air coming from the attacked leaf started to emit the same gases themselves, and they also produced nectar, while the leaves that were not exposed to the air from the attacked leaf remained the same.

The results were significant because they revealed that the gases emitted from an attacked leaf are necessary for the same plant to protect *its other leaves* from future attacks. In other words, when a leaf is attacked by an insect or by bacteria, it releases odors that warn its brother leaves to protect themselves against imminent attack, similar to guard towers on the Great Wall of China lighting fires to warn of an oncoming assault. In this way, the plant ensures its own survival as leaves that have "smelled" the gases given off by the attacked leaves will be more resistant to the impending onslaught.

So what about the neighboring plant? If it's close enough to the attacked plant, then it benefits from this internal "conversation" among leaves on the infested plant. The neighboring plant eavesdrops on a nearby olfactory conversation, which gives it essential information to help protect itself. In nature, this olfactory signal persists for at least a few feet (different volatile signals, depending on their chemical properties, travel for shorter or much longer distances). For lima beans, which naturally enjoy crowding, this is more than enough to ensure that if one plant is in trouble, its neighbors will know about it.

What exactly is the lima bean smelling when its neighbor is eaten? *Eau de lima*, just like the *eau de tomato* described in the dodder experiment, is a complex mixture of aromas. In 2009, Heil collaborated with colleagues from South Korea and analyzed the different volatile compounds emitted from the leaves of the attacked plants in order to identify the chemical messenger. But the trick was identifying the one chemical responsible for the evident communication with other leaves. They compared the compounds emitted by leaves following bacterial infection with those emitted following an insect feeding. Both treatments resulted in the expression of similar volatile gases, except for two gases that discriminated between the two treatments. The leaves under *bacterial* attack emitted a gas called methyl salicylate, and those eaten by bugs did not; the latter produced a gas called methyl jasmonate.

Methyl salicylate is very similar in structure to salicylic acid. Salicylic acid is found in copious amounts in the bark of willow trees. In fact, the ancient Greek physician Hippocrates described a bitter substance, now known to be salicylic acid, from willow bark that could ease aches and reduce fevers. Other cultures in the ancient Middle East also used willow bark as a medicine, as did Native Americans. Centuries later, we know salicylic acid as the chemical precursor for aspirin (which is acetylsalicylic acid), and salicylic acid itself is a key ingredient in many modern anti-acne face washes.

Although willow is a well-known producer of salicylic acid, which has been extracted from this tree for years, all plants produce this chemical in various amounts. They also produce methyl salicylate (which by the way is an important ingredient in Bengay ointment). But why would a plant produce a pain reliever and fever reducer? As with any phytochemical (or plant-produced chemical), plants don't make salicylic acid for *our* benefit. For plants, salicylic acid is a "defense hormone" that potentiates the plant's immune system. Plants produce it when they've been attacked by bacteria or viruses. Salicylic acid is soluble and released at the exact spot of infection to signal through the veins to the rest of the plant that bacteria are on the loose. The healthy parts of the plant respond by initiating a number of steps that either kill the bacteria or, at the very least, stop the plague's spread. Some of these include putting up a barrier of dead cells around the site of infection, which blocks the movement of the bacteria to other parts of the plant. You sometimes see these barriers on leaves; they appear as white spots. These spots are areas of the leaf where cells have literally killed themselves so that the bacteria near them can't spread farther. *Good idea for a poem*

At a broad level, salicylic acid serves similar functions in both plants and people. Plants use salicylic acid to help ward off infection (in other words, when they're sick). We've used salicylic acid since ancient times, and we use the modern derivative of aspirin when we're sick with an infection that causes aches and pains.

Returning to Heil's experiments: his lima beans emitted methyl salicylate, a volatile form of salicylic acid, after they were attacked with bacteria. This result supported work done a decade earlier in the lab of Ilya Raskin at Rutgers University, who had shown that methyl salicylate was the major volatile compound produced by tobacco following viral infection. Plants can convert soluble salicylic acid to volatile methyl salicylate and vice versa. One way to understand the difference between salicylic acid and methyl salicylate is this: plants *taste* salicylic acid, and they *smell* methyl salicylate. (As we know, taste and smell are interrelated senses. The major difference is that we *taste* soluble molecules on the tongue, while we *smell* volatile molecules in the nose.)

By enclosing the infected leaves in plastic bags, Heil had blocked the methyl salicylate from wafting from the infected leaf to the noninfected one, whether on the same vine or on a neighboring plant. When the noninfected leaf finally smelled the methyl salicylate by having the air from the infected leaf blown onto it, it inhaled the gases through the tiny openings on the leaf surface (called stomata). Once deep in the leaf, the methyl salicylate was converted back to salicylic acid, which, as we now know, plants take when they're feeling sick.*

---

*If you're wondering about methyl jasmonate, the story is quite similar. Methyl jasmonate is a voluble form of jasmonic acid, a defense hormone that plants emit upon leaf damage inflicted by herbivorous creatures.

Do Plants Smell?

Plants give off a literal bouquet of smells. Imagine the smell of roses when you walk on a garden path in the summertime, or of freshly cut grass in the late spring, or of jasmine blooming at night. How about the sweet pungent odor of a brown banana intemingled with the myriad of smells at a farmers' market? Without looking, we know when fruit is ready to eat, and no visitor to a botanical garden can be oblivious to the offensive odor of the world's largest (and smelliest) flower, the *Amorphophallus titanum*, better known as the corpse flower. (Luckily, it blooms only once every few years.)

Many of these aromas are used in complex communication between plants and animals. The smells induce different pollinators to visit flowers and seed spreaders to visit fruits, and as

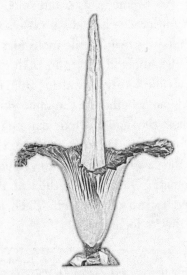

Corpse flower (*Amorphophallus titanum*)

the author Michael Pollan infers, these aromas can even induce people to spread flowers all over the world. But plants don't just give off odors; as we've seen, they undoubtedly smell other plants.

Of course, like plants, we sense airborne volatile compounds. We use our noses to get a whiff of many things, particularly food. But we need to remember that "olfaction" means much more than smelling good food. Our language is littered with olfactory-tainted statements like "the smell of fear" and "I smell trouble," and smells are intimately tied with memory and emotion. The olfactory receptors in our noses are directly connected to the limbic system (the control center for emotion) and evolutionarily to the most ancient part of our brains. Like plants, we communicate via pheromones, though we're often not aware of it.

Pheromones are given off by one individual and trigger a social response in another. Pheromones in different animals, from flies to baboons, communicate various situations: social dominance, sexual receptiveness, fear, and so on. We also are influenced by odors and emit odors that affect those around us. For example, synchronization of menstrual cycles among women living in close quarters has been found to be due to odor cues in perspiration. A recent (and provocative) study in *Science* showed that men who simply sniffed negative-emotion-related odorless tears obtained from women showed reduced levels of testosterone and reductions in sexual arousal. So subtle olfactory signals could potentially affect many aspects of our psyche.

Plants and animals sense volatile compounds in the air, but can this really be considered olfaction by plants? Plants obviously don't have olfactory nerves that connect to a brain that interprets the signals, and as of 2011 only one receptor for a volatile chemical, the ethylene receptor, has been identified in plants. But ripening fruits, *Cuscuta*, Heil's plants, and other flora throughout our natural world respond to pheromones, just as we do. Plants detect a volatile chemical in the air, and they convert this signal (albeit nerve-free) into a physiological response. Surely this could be considered olfaction.

So if plants can "smell" in their own unique ways without olfactory nerves, is it possible they can "feel" as well without *sensory* nerves?

THREE

What a Plant Feels

I will touch a hundred flowers
And not pick one.

    —Edna St. Vincent Millay, "Afternoon on a Hill"

Most of us interact with plants every day. At times we experience plants as soft and comforting, like grass in a park during an indulgent midday nap or fresh rose petals spread across silk sheets. Other times they are rough and prickly: we navigate around pesky thorns to get to a blackberry bush on a meander through the woods or trip over a knotted tree trunk that's worked its way up through the street. But in most cases, plants remain passive objects, inert props that we interact with but ignore while we do so. We pluck petals from daisies. We saw the limbs off unsightly branches. What if plants knew we were touching them?

It's probably a bit surprising, and maybe even a bit disconcerting, to discover that plants know when they're being touched. Not only do they know when they're being touched, but plants can differentiate between hot and cold, and know when their branches are swaying in the wind. Plants feel direct contact: some plants, like vines, immediately start rapid growth upon contact with an object like a fence they can wrap themselves around, and the Venus flytrap purposely snaps its jaws shut when an insect lands on its leaves. And plants seemingly don't like to be touched too much, as simply touching or shaking a plant can lead to growth arrest.

Of course, plants don't "feel" in the traditional sense of the term. Plants don't feel regret; they don't get a feel for a new job. They do not have an intuitive awareness of a mental or emotional state. But plants perceive tactile sensation, and some of them actually "feel" better than we do. Plants like the burr cucumber (*Sicyos angulatus*) are up to ten times more sensitive than we are when it comes to touch. Vines from a burr cucumber can feel a string weighing only 0.009 ounce (0.25 gram), which is enough to induce the vine to start winding itself around a nearby object. On the other hand, most of us can only feel the presence of a very light piece of string on our fingers that is about 0.07 ounce (or 2 grams) in weight. Although a plant may be more sensitive to touch than a human being, plants and animals have some surprising similarities when it comes to feeling that touch.

Our sense of touch transmits vastly different sensations, from a painful burn to the light trace of a breeze. When we come into contact with an object, nerves are activated, sending a signal to the brain that communicates the type of sensation—pressure, pain, temperature, and more. All physical sensations are perceived through our nervous system by specific sensory neurons in our skin, muscles, bones, joints, and internal organs. Through the action of different types of sensory neurons, we experience a vast array of physical sensations: tickling, sharp pain, heat, light touch, or dull ache, to name a few. Just as different types of pho-

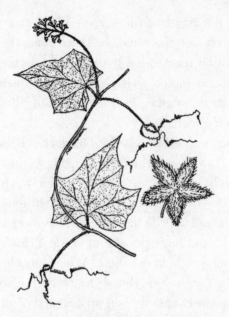

Burr cucumber (*Sicyos angulatus*)

toreceptors are specific for different colors of light, different sensory neurons are specific for different tactile experiences. Different receptors are activated by an ant crawling on your arm and by a deep Swedish massage at the spa. Our bodies have receptors for cold and receptors for heat. But each of the different types of sensory neurons acts in essentially the same way. When you touch something with your fingers, the sensory neurons for touch (known as mechanoreceptors) relay their signal to an intermediary neuron that connects up to the central nervous system in the spinal cord. From there, other neurons convey the signal to the brain, which tells us that we've felt something.

The principle involved in neural communication is the same for all nerve cells: electricity. The initial stimulus starts a rapid electrochemical reaction known as depolarization, which is propagated along the length of the nerve. This electric wave hits the adjacent neuron, and the wave continues along the new neuron, and so on, until it reaches the brain. A block in the signal at any stage can be catastrophic, as in the case of traumatic spine damage, which cuts this signal, leading to loss of feeling in the limbs that have been affected.

While the mechanisms involved in electrochemical signaling are complex, the basic principles are simple. Just as a battery maintains its charge by housing different electrolytes in different compartments, a cell has a charge owing to different amounts of various salts in and outside the cell. There is more sodium on the outside of cells and more potassium inside. (That's why salt balances are so important in our diets.) When a mechanoreceptor is activated, let's say by your thumb touching the space bar on a keyboard, specific channels open up near the point of contact in the cell membrane that allow sodium to pass into the cell. This movement of sodium changes the electric charge, which leads to the opening of additional channels and increased sodium flux. This results in the depolarization that propagates along the length of the neuron like a wave propagating across the ocean.

At the end of a neuron, at the junction where it meets an adjacent one, the action potential leads to a rapid change in the concentration of an additional ion, calcium. This calcium spike is necessary for the release of neurotransmitters from the active neuron, which are received by the next neuron. Neurotransmitters binding to the second neuron initiate new waves of action potentials. These spikes in electrical activity exemplify the ways in which nerves communicate, whether from a receptor to the brain or from the brain to a muscle to cause a movement. The ubiquitous cardiac monitors in hospitals depict this kind of electrical activity as it relates to heart function—a spike of activity followed by a recovery, which is repeated again and again. Mechanosensory neurons send similar spikes of activity to the brain, and the frequency of the spikes communicates the strength of the sensation.

But touch and pain are biologically not the same phenomena. Pain does not simply result from an increase in the signals emanating from touch receptors. Our skin features distinct receptor neurons for different types of touch, but it also has unique receptor neurons for different types of pain. Pain receptors (called nociceptors) require a much stronger stimulus before they send action potentials to the brain. Advil, Tylenol, and other pain relievers work because they specifically mute the signal coming from the nociceptors but not the mechanoreceptors.

So human touch is a combination of actions in two distinct parts of the body—cells that sense the pressure and turn this pressure into an electrochemical signal, and the brain that processes this electrochemical signal into specific types of feelings and initiates a response. So what happens in plants? Do they have mechanoreceptors?

Venus's Trap

The Venus flytrap* (otherwise known as *Dionaea muscipula*) is the quintessential example of a plant that responds to touch. It grows in the bogs of the Carolinas, where the soil lacks nitrogen

---

*The "Venus" part of the plant's name has little to do with science and much to do with the rather lewd imaginations of nineteenth-century English botanists. See www.sarracenia.com/faq/faq2880.html.

and phosphorous. To survive in such a nutritionally poor environment, *Dionaea* has evolved the amazing ability to garner nutrition not only from light but from insects—and small animals as well. These plants carry out photosynthesis, like all green plants, but they moonlight as carnivores, supplementing their diet with animal protein.

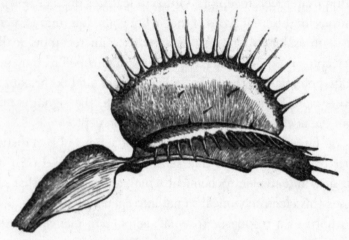

Venus flytrap (*Dionaea muscipula*)

The leaves of the Venus flytrap are unmistakable: they end in two main lobes connected by a central midrib, and the edges of the two lobes are bordered by long protrusions, called cilia, that resemble the teeth of a comb. These two lobes, connected on one side by a hinge, are normally spread at an angle, forming a V-like structure. The internal sides of the lobe have pink and purple hues and excrete nectar that is irresistible to many creatures. When an unassuming fly, a curious beetle, or even a small, meandering frog crawls across the surface of the leaves, the two leaves spring together with surprising force, sandwiching the unsuspecting prey and blocking its escape with its jail bars of interlocking cilia.* The trap closes at an astounding speed: unlike our futile attempts to swat at a pesky fly, the Venus flytrap springs shut in less than one-tenth of a second. Once activated, the trap excretes digestive juices that dissolve and absorb the poor prey.

The amazing characteristics of the Venus flytrap led Charles Darwin, who was among the first scientists to publish an in-depth study of the plant and other carnivorous flora, to describe it as "one of the most wonderful [plants] in the world." Darwin's interest in carnivorous plants illustrates how naive curiosity can lead a trained scientist to groundbreaking discoveries. Darwin begins his 1875 treatise *Insectivorous Plants* in this way: "During the summer of 1860, I was surprised by finding how large a number of insects were caught by the leaves of the common sun-dew [plant] (*Drosera rotundifolia*) on a heath in Sussex. I had heard that insects were thus caught, but knew nothing further on the subject." From knowing virtually nothing about the matter, Darwin became the foremost expert on carnivorous plants, including the Venus flytrap, in the nineteenth century, and indeed his work is still referenced today.

We now know that the Venus flytrap feels its prey and senses if the organism crawling around inside its trap is the right size to consume. There are several large black hairs on the pink surface of the inside of each lobe, and the hairs act as triggers that spring the trap closed. But one hair being touched is not enough to

---

*See www.youtube.com/watch?v=ymnLpQNyI6g for a great example of the Venus flytrap in action.

spring the trap; studies have revealed that at least two have to be touched within about twenty seconds of each other. This ensures that the prey is the ideal size and won't be able to wiggle out of the trap once it closes. The hairs are extremely sensitive, but they are also very selective. As Darwin noted in his book *Insectivorous Plants*:

> Drops of water, or a thin broken stream, falling from a height on the filaments [hairs], did not cause the blades to close . . . No doubt, the plant is indifferent to the heaviest shower of rain . . . I blew many times through a fine pointed tube with my utmost force against the filaments without any effect; such blowing being received with as much indifference as no doubt is a heavy gale of wind. We thus see that the sensitiveness of the filaments is of a specialized nature.

Even though Darwin described in great detail the series of events leading to trap closure and the nutritional advantage of the animal protein to the plants, he couldn't come up with the mechanism of the signal that differentiated between rain and fly and enabled the rapid imprisonment of the latter. Convinced that the leaf was absorbing some meaty flavor from the prey on the lobes, Darwin tested all types of proteins and substances on the leaf. But these studies were for naught, as he could not induce trap closure with any of his treatments.

His contemporary John Burdon-Sanderson made the crucial discovery that explained the triggering mechanism once and for all. Burdon-Sanderson, a professor of practical physiology at University College in London and a physician by training, studied the electrical impulses found in all animals, from frogs to mammals, but from his correspondence with Darwin became particularly fascinated by the Venus flytrap. Burdon-Sanderson carefully placed an electrode in the Venus flytrap leaf, and he discovered that pushing on two hairs released an action potential very similar to those he observed when animal muscles contract. He found that it took several seconds for the electrical current to return to its resting state after it had been initiated. He realized that when an insect brushes up against the hairs inside the trap, it induces a depolarization that is detected in both lobes.

Burdon-Sanderson's discovery that pressure on two hairs leads to an electrical signal that is followed by the trap closing was one of the most important of his career and was the first demonstration that electrical activity regulates plant development. But he could only hypothesize that the electric signal was the direct cause of trap closure. More than one hundred years later, Alexander Volkov and his colleagues at Oakwood University in Alabama proved that the electric stimulation itself is the causative signal for the trap's closing. They applied a form of electric shock therapy to the open lobes of the plant, and it caused the trap to close without any direct touch to the trigger hairs. Volkov's work and earlier research in other labs also made it clear that the trap remembers if only a single trigger hair has been touched, and then it waits until a second hair is triggered before closing. Only very recently did this research shed light on the mechanism that allows the Venus flytrap to remember how many of its hairs have been triggered, which I'll explore in chapter six. Before we get to the ways in which plants remember, we need to take some time with the connection between the electric signal and the movement of the leaves.

Water Power

Burdon-Sanderson observed that the electrical pulse he detected in the closing Venus flytrap was very similar to the action of a nerve and a contracting muscle. While the action potentials in the absence of nerves were clear to him, the mechanism of movement in the absence of muscles was obscure. To Burdon-Sanderson's knowledge, the action potential in the plant had no clear muscle-like target to act on to induce the trap's closing.

Studies of *Mimosa pudica* provided a wonderful experimental system to understand the world of leaf movements, which could then be generalized to other plants. The *Mimosa pudica* is native to South and Central America but is now grown worldwide as an ornamental because of its fascinating moving leaves. Its leaves are hypersensitive to touch, and if you run your finger down one of them, all the leaflets rapidly fold inward and droop. They reopen several minutes later, only to rapidly close once more if you touch them again. The name *pudica* reflects this drooping movement. It means "shy" in Latin. The plant is also known throughout many regions as "the sensitive plant." Its unusual behavior is referred to as "false death" in the West Indies, and it is referred to as the "don't touch me" plant in Hebrew and the "shy virgin" in Bengali.

The *Mimosa*'s characteristic drooping and opening action is very similar to that of the Venus flytrap even at the level of electrophysiology. This was noticed by Sir Jagadish Chandra Bose, a noted physicist turned plant physiologist from Calcutta, India. While carrying out research in the Davy Faraday Research Laboratory of the Royal Institution of Great Britain, Bose reported to the Royal Society in a lecture in 1901 that touch initiated an electric action potential that radiated the length of the leaf,

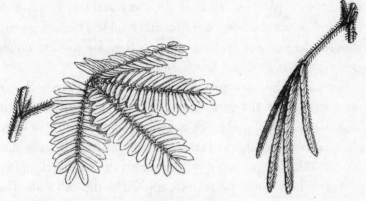

Mimosa pudica

resulting in the rapid closing of the *Mimosa* leaflets. (Unfortunately, Burdon-Sanderson was highly critical of Bose's work and recommended that his *Mimosa* paper be rejected from the *Proceedings of the Royal Society of London*, though subsequent studies in many labs have since shown that Bose was indeed correct.)

Studies revealed that when the electric signal acts on a group of cells called the pulvinus, which are the motor cells that move the leaves, it leads to the drooping behavior of the *Mimosa's* leaves. To understand how the pulvinus moves the leaves in the absence of muscles, we have to understand a little bit of basic plant cell biology. The plant cell contains two main parts. The protoplast, similar to cells in animals, resembles a water balloon: a thin membrane surrounds a liquid interior. This interior contains several microscopic parts, including the nucleus, mitochondria, proteins, and DNA. What's unique about plant cells is that the protoplast is enclosed within the second part, a boxlike structure called the cell wall. The cell wall gives a plant its strength in the absence of a supporting skeleton. In wood, cotton, and nutshells, for example, the cell walls are thick and sturdy, while in leaves and petals the walls are thin and pliable. (We are incredibly dependent on cell walls, in fact, as they are used to create paper, furniture, clothing, ropes, and even fuel.)

Normally, the protoplast contains so much water that it presses strongly on the surrounding cell wall, which allows plant cells to be very tight and erect and to support weight. But when a plant lacks water, there's little pressure on the cell walls, and the plant wilts. By pumping water in and out of cells, the plant can control how much pressure is applied to the cell wall. The pulvinus cells are found at the base of each *Mimosa* leaflet and act as mini hydraulic pumps that move the leaves. When the pulvinus cells are filled with water, they push the leaflets open; when they lose water, the pressure drops, and the leaves fold into themselves.

Where do the electric action potentials come into play? They are the critical signal that tells the cell whether to pump water in or out. Under normal conditions, when the *Mimosa*'s leaves are open, the pulvinus cells are full of potassium ions. The high concentration of potassium inside the cell relative to the outside causes water to enter the cell in a futile attempt to dilute the potassium, which results in great pressure on the cell wall—and in erect leaves. Potassium channels are opened when the electric signal reaches the pulvinus, and as the potassium leaves the cell, so does the water. This causes the cells to become flaccid. Once the signal has passed, the pulvinus pumps potassium into the cells again, and the resulting influx of water opens up the leaf again. Calcium, the same ion critical for neural communication in humans, regulates the opening of the potassium channels, and as we'll see, it is essential for a plant's response to touch.

A Negative Touch

In the early 1960s, Frank Salisbury was studying the chemicals that induce flowering in cocklebur (*Xanthium strumarium*), a weed found throughout North America and most notorious for its little football-shaped burrs, which are commonly found clinging to hikers' clothing. To understand how the plant grew, Salisbury and his team of technicians at Colorado State University decided to measure the daily increase in leaf length by going out into the field and physically measuring the leaves with a ruler. To his bewilderment, Salisbury noticed that the leaves being measured never reached their normal length. Not only that, as the experiment continued, they eventually turned yellow and died. But the leaves on the same plant that were not handled and measured thrived. As Salisbury explained, "We were confronted with

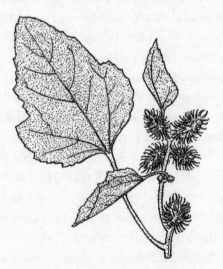

Cocklebur (*Xanthium strumarium*)

the remarkable discovery that one can kill a cocklebur leaf simply by touching it for a few seconds each day!"

As Salisbury's interests lay elsewhere, a decade passed before his observation was put into broader context. Mark Jaffe, a plant physiologist who was based at Ohio University in the early 1970s, recognized that this touch-induced growth inhibition is a general phenomenon in plant biology. He coined the cumbersome term "thigmomorphogenesis" from the Greek roots *thigmo-* (touch) and *morphogenesis* (creation of shape) to describe the general effect of mechanical stimulation on plant growth.

Of course, plants are exposed to multiple tactile stresses such as wind, rain, and snow, and animals regularly come into contact with many of them. So in retrospect, it isn't so surprising that a plant would retard its growth in response to touch. A plant feels what type of environment it lives in. Trees growing high on a mountain ridge are often exposed to strong winds, and they adapt to this environmental stress by limiting their branch development and growing short, thick trunks. The same species of tree grown in a protected valley, on the other hand, will be tall, thin, and full of branches. Growth retardation in response to touch is an evolutionary adaptation that increases the chances that a plant can survive multiple, and often violent, perturbations. Indeed, from an ecological point of view, a plant faces many of the same choices that we would if we were building a home. What types of resources should go into the foundation? How about the frame? If you live in an area with low wind levels, or low risk for earthquakes, then your resources may go into the outward appearance of your home. But in an area of high wind levels, or high risk for earthquakes, your resources have to go into a substantial foundation and frame.

What holds true for trees also holds true for our little mustard plant *Arabidopsis thaliana* that we met in the first chapter. An arabidopsis plant that's touched a few times a day in the lab will be much squatter, and flower much later, than one that's left to its own accord. Simply stroking its leaves three times a day completely changes its physical development. While this change in overall growth takes many days for us to witness, the initial cellular response is actually quite rapid. In fact, Janet Braam and her colleagues at Rice University demonstrated that simply touching an arabidopsis leaf results in a rapid change in the *genetic* makeup of the plant.

That Braam discovered this phenomenon at all was quite serendipitous. Initially, as a young research fellow at Stanford University, she was interested not in the effect of touch on plants but rather in the genetic programs activated by plant hormones. In one of her experiments designed to elucidate the effect of the hormone gibberellin on plant biology, she sprayed arabidopsis leaves with this hormone and then checked which genes were activated by the treatment. She discovered several genes that were rapidly turned on following her spray treatment, and she assumed that they were responding to the gibberellin. But it turned out that their activity increased after they were sprayed with any number of substances—even water.

Not to be defeated, Braam pressed on, trying to figure out why these genes were activated even by water. She experienced a true eureka moment when she realized that the common factor in the treatments was the *physical sensation* of being sprayed with the solutions. Braam hypothesized that the genes she discovered were responding to the physical treatment of the leaves. To test this, she continued her experiment, but rather than spraying the plants with water, she simply touched them. To her satisfaction, the same genes that had been induced by spraying with the hormone, or with water, were also activated by touching the plants. Braam understood that her newfound genes were clearly activated by touch, and she aptly named them the "*TCH* genes" since they were induced by touching the plants.

Further understanding the importance of this discovery necessitates a quick exploration of how genes work in general. The DNA found in the nucleus of each cell that makes up an arabidopsis plant contains about twenty-five thousand genes. At the simplest level, each gene encodes one protein. While the DNA is the same in each cell, different cells contain different proteins. For example, a cell in a leaf contains proteins different from those in a root cell. The leaf cell contains proteins that absorb light for photosynthesis, while the root cell contains proteins that help it absorb minerals from the soil. Various cell types contain different proteins because different genes are active—or more exactly, different genes are transcribed—in each type of cell. While some genes are transcribed in all cells (like the genes needed for making membranes, for example), most genes are transcribed in only specific subsets of cell types. So while each cell has the *potential* to turn on any of the twenty-five thousand genes, in practice only several thousand genes are active in a particular cell type. To further complicate matters, many genes are also controlled by the external environment. Some genes are transcribed in leaves only after the leaves see blue light. Some are transcribed in the middle of the night, some after a heat spell, some after a bacterial attack, and some after touch.

What are these touch-activated genes? The first of the *TCH* genes Braam identified encode proteins involved in calcium signaling in the cell. As we've seen earlier, calcium is one of the important salt ions that regulates both the cell's electrical charge and communication between cells. In plant cells, calcium helps maintain cell turgor (as in the pulvinus cells in the case of the *Mimosa* plant) and is also part of the plant cell wall. Calcium is essential for humans and other animals to propagate electric signals from neuron to neuron, and it is also necessary for muscle contraction. Although we do not yet know all of the ways in which calcium regulates such diverse phenomena at the same time, it is a field of intense study.

Scientists do know that following a mechanical stimulation like the shaking of a branch or a root hitting a rock, the concentration of calcium ions in a plant cell peaks rapidly and then drops. This spike affects the charge across the cell membrane, but it also directly affects multiple cellular functions as a "second messenger," a mediator molecule that relays information from specific receptors to specific outputs. This free soluble calcium is not very efficient on its own in causing some response, because most proteins can't bind calcium directly; hence calcium, in both plants and animals, usually works in conjunction with a small number of calcium-binding proteins.

Among these, the most studied is calmodulin (*cal*cium-*modul*ated prote*in*). Calmodulin is a relatively small but very important protein, and when it binds with calcium, it interacts with, and modulates the activity of, a number of proteins involved in processes in human beings—such as memory, inflammation, muscle function, and nerve growth. Getting back to plants, Braam showed that the first *TCH* gene encoded calmodulin. In other words, when you touch a plant, be it arabidopsis or papaya, one of the first things it does is make more calmodulin. Most likely, a plant makes more calmodulin to work with the calcium that it releases during the action potentials.

Thanks to the continuing work of Braam and other scientists, we now know that over 2 percent of arabidopsis genes (including, but not limited to, genes encoding calmodulin and other calcium-related proteins) are activated after an insect lands on its leaf, an animal brushes up against it, or the wind moves its branches. This is a surprisingly large number of genes, which indicates just how far-reaching a plant's response is when it comes to mechanical stimulation and survival.

Plant and Human Feeling

We can feel a varied and complex mixture of physical sensations due to the presence of specialized mechanosensory receptor nerves and due to a brain that translates these signals into sensations with emotional connotations. These receptors enable us to respond to a vast array of tactile stimulations. A specific mechanosensory receptor called Merkel's disks detects sustained touch and pressure on our skin and muscles. Nociceptors in our mouths are activated by capsaicin, the ultrahot chemical found in chili peppers, and nociceptors signal that our appendix is inflamed before an appendectomy. Pain receptors exist to enable us to withdraw from a dangerous situation or to let us know of a potentially dangerous physical problem inside our bodies.

While plants feel touch, they don't feel pain. Their responses are also not subjective. Our perception of touch and pain is subjective, varying from person to person. A light touch can be pleasurable to one person or an annoying tickle to another. The basis for this subjectivity ranges from genetic differences affecting the threshold pressure needed to open an ion channel to psychological differences that connect tactile sensations with associations such as fear, panic, and sadness, which can exacerbate our physiological reactions.

A plant is free from these subjective constraints because it lacks a brain. But plants feel mechanical stimulation, and they can respond to different types of stimulation in unique ways. These responses do not help the plant avoid pain but modulate development to best suit the ambient environment. An amazing example of this was provided by Dianna Bowles and her team of researchers at the University of Leeds. Earlier research had

Tomato (*Solanum lycopersicum*)

shown that wounding a single tomato leaf leads to responses in the unwounded leaves on the same plant (similar to the types of research outlined in chapter two). These responses include the transcription of a class of genes called proteinase inhibitors in the intact leaves.

Bowles was curious to know more about the nature of the signal from a wounded leaf to an unharmed leaf. The accepted paradigm was that a secreted chemical signal was transported in the veins of a wounded leaf to the rest of the plant. But Bowles hypothesized that the signal was electric. To test her hypothesis, she burned one tomato leaf with a hot steel block and found that an electric signal could be detected in the stem of the same plant at a distance from the wounded leaf. The plant could still detect the signal even if she iced the stem-like structure that connects the leaf to the stem (called a petiole). She found that icing the petiole blocked chemical flow from the leaf to the stem—but not electrical flow. Moreover, when she iced the petiole of the burned leaf, the untreated leaves still transcribed the proteinase inhibitor genes. The leaf did not feel pain. The tomato responded to the hot metal not by moving away from it but by warning its other leaves of a potentially dangerous environment.

As sessile, rooted organisms, plants may not be able to retreat or escape, but they *can* change their metabolism to adapt to different environments. Despite the differences between the ways plants and animals react to touch and other physical stimulations at the organismal level, at the cellular level the signals initiated are hauntingly similar. Mechanical stimulation of a plant cell, like mechanical stimulation of a nerve, initiates a cellular change in ionic conditions that results in an electric signal. And just like in animals, this signal can propagate from cell to cell, and it involves the coordinated function of ion channels including potassium, calcium, calmodulin, and other plant components.

A specialized form of mechanoreceptor is also found in our ears. So if plants can sense touch as a result of the mechanoreceptors similar to the ones in our skin, can they also hear by sensing sound through mechanoreceptors similar to the ones in our ears?

FOUR

What a Plant Hears

The temple bell stops but I still hear the sound coming
out of the flowers.

—Matsuo Bashō

Forests reverberate with sounds. Birds sing, frogs croak, crickets
chirp, leaves rustle in the wind. This never-ending orchestra in-
cludes sounds that signal danger, sounds related to mating rituals, sounds that threaten, sounds that appease. A squirrel jumps
on a tree at the crunch of a breaking branch; a bird answers the
call of another. Animals constantly move in response to sound,
and as they move, they create new sounds, contributing to a
cyclical cacophony. But even as the forest chatters and crackles,
plants remain ever stoic, unresponsive to the din around them.
Are plants deaf to the clamor of a forest? Or are we just blind to
their response?

While various forms of rigorous scientific research have
helped shed light on the plant senses we've covered so far, little
credible, conclusive research exists when it comes to a plant's response to sound. This is surprising, given the amount of anec-
dotal information we have about the ways in which music may

influence how a plant grows. While we may think twice when we hear that plants can smell, the idea that plants can hear comes as no surprise at all. Many of us have heard stories about plants flourishing in rooms with classical music (though some people claim it's really pop music that gets a plant moving). Typically, though, much of the research on music and plants has been carried out by elementary school students and amateur investigators who do not necessarily adhere to the controls found in laboratories grounded in the scientific method.

Before we delve into whether or not plants can actually hear, let's get a better understanding of human hearing. A common definition of "hearing" is "the ability to perceive sound by detecting vibrations via an organ such as the ear." Sound is a continuum of pressure waves that propagate through the air, through water, or even through solid objects such as a door or the earth. These pressure waves are initiated by striking something (such as by beating a drum) or starting a repeated vibration (like plucking a string) that causes the air to compress in a rhythmic fashion. We sense these waves of air pressure through a particular form of mechanoreception by tactile-sensitive hair cells in our inner ears. These hair cells are specialized mechanosensory nerves from which extend hairlike filaments called stereocilia that bend when an air-pressure wave (a sound) hits them.

The hair cells in our ears convey two types of information: volume and pitch. Volume (in other words, the *strength* of the sound) is determined by the height of the wave reaching the ear, or what's better known as the amplitude of the waves. Loud noises have high amplitude, and soft noises have low amplitude. The higher the amplitude, the more the stereocilia bend. Pitch, on the other hand, is a function of the *frequency* of the pressure waves—how many times per second the wave is detected regardless of its amplitude. The faster the frequency of the wave, the faster the stereocilia bend back and forth and the higher the pitch.*

As the stereocilia vibrate in the hair cells, they initiate action potentials (as do other types of mechanoreceptors that we encountered in the previous chapter) that are relayed to the auditory nerve, and from there they travel to the brain, which translates this information into different sounds. So human hearing is the result of two anatomical occurrences: the hair cells in our ears receive the sound waves, and our brain processes this information so that we can respond to different sounds. Now, if plants are capable of detecting light without having eyes, can they detect sound if they don't have ears?

Rock-and-Roll Botany

At one point or another, many of us have been intrigued by the idea that plants respond to music. Even Charles Darwin (who, as we've seen, carried out his seminal research into plant vision and feeling over a century ago) studied whether plants could pick up on the tunes he played for them. In one of his more bizarre experiments, Darwin (who, in addition to his lifetime commit-

---

*Sound waves are measured in hertz (Hz), where 1 Hz equals one wave cycle per second. We can hear sound waves in the range of 20 Hz for low pitches to up to 20,000 Hz for the highest pitches. The lowest note on a contrabass, for example (the low E), vibrates at 41.2 Hz, while the highest note on a violin (high E) vibrates at 2,637 Hz. The highest C on a piano vibrates at 4,186 Hz, and the C two octaves above this vibrates at about 16,000 Hz. A dog's ear responds to sound waves above 20,000 Hz (which is why we can't hear dog whistles), and bats even emit and detect rebounding sound waves up to 100,000 Hz for their internal sonar that maps out the landscape ahead of them. At the other end of the spectrum, an elephant can hear and vocalize sounds *below* 20 Hz, which human beings also can't detect.

ment to biology research, was an avid bassoonist) monitored the effects of his own bassoon music on plant growth by seeing if his bassoon could induce the leaves of the *Mimosa* plant to close (it couldn't, and he described his study as a "fool's experiment").

Research dealing with plant auditory prowess hasn't exactly blossomed since Darwin's failed attempts. Hundreds of scientific articles have been published in the last year alone that deal with plant responses to light, smell, and touch, yet only a handful have been published over the last twenty years that have dealt specifically with plant responses to sound, and even then many of them don't hold up to my standards for what would be evidence of a "hearing" plant.

An example of one of these papers (albeit a zany one) was published in *The Journal of Alternative and Complementary Medicine*. It was written by Gary Schwartz, a professor of psychology and medicine, and his colleague Katherine Creath, a professor of optical sciences, both based at the University of Arizona, where Schwartz founded the VERITAS Research Program. This program "test[s] the hypothesis that the consciousness (or personality or identity) of a person survives physical death." Obviously, studying consciousness after death presents some experimental difficulties, so Schwartz also studies the existence of "healing energy." Because human participants in a study can be strongly influenced by the power of suggestion, Schwartz and Creath used plants instead, in order to uncover the "biologic effects of music, noise, and healing energy." Of course, plants cannot be influenced by the placebo effect or, as far as we know, by musical preferences (though researchers carrying out and analyzing experiments can be).

They hypothesized that healing energy and "gentle" music (which consisted of Native American flute and nature sounds, which they noted were preferred by the experimenter) would be conducive to the germination of seeds.* Creath and Schwartz explained that their data revealed that slightly more zucchini and okra seeds germinated in the presence of gentle music sounds than seeds that were kept in silence. They also noted that germination rates could increase as a result of Creath's healing energy, which she applied to the seeds with her hands.† It goes without saying that these results have not been validated by subsequent research in other plant laboratories, but one of the sources Creath and Schwartz cited in support of their results was Dorothy Retallack's *The Sound of Music and Plants*.

Dorothy Retallack was a self-described "doctor's wife, housekeeper and grandmother to fifteen," and she enrolled as a freshman in 1964 in the now-defunct Temple Buell College after her last child had graduated from college. Retallack, a professional mezzo-soprano who often performed at synagogues, churches, and funeral homes, decided to major in music at Temple Buell. She took an Introduction to Biology course to complete her science requirements and was asked by her teacher to carry out any experiment that she thought would interest her. Retallack's juxtaposing of her biology requirement with her love for music resulted in a book spurned by mainstream science but quickly embraced by the popular culture.

Retallack's *The Sound of Music and Plants* provides a window into the cultural-political climate of the 1960s, but it also

---

*It's interesting that they chose "gentle" sounds as they cite Pearl Weinberger from the University of Ottowa, who used ultrasonic waves (which are definitely not gentle) in her studies in the 1960s and 1970s.

†Creath was trained in VortexHealing, which is described as "a Divine healing art and path for awakening. It is designed to transform the roots of emotional consciousness, heal the physical body, and awaken freedom within the human heart. This is the Merlin lineage." See www.vortexhealing.com.

sheds light on her perspective as well. Retallack comes across as a unique mixture of a social conservative who believed that loud rock music correlated with antisocial behavior among college students and a New Age religious spiritualist who saw a sacred harmony between music and physics and all of nature.

Dorothy Retallack in the lab with her adviser Dr. Francis Broman

Retallack explained that she was intrigued by a book published in 1959 titled *The Power of Prayer on Plants*, in which the author claimed that plants that were prayed to thrived while those bombarded with hateful thoughts died. Retallack wondered whether similar effects could be induced by positive or negative genres of music (the ruling of positive or negative was

dictated, of course, by her own musical taste). This question became the basis of her research requirement. By monitoring the
effect of different music genres on plant growth, she hoped to
provide her contemporaries with proof that rock music was potentially harmful—not only to plants, but to humans as well.

Retallack exposed different plants (philodendrons, corn,
geraniums, and violets, to name a few—each experiment used a
different species) to an eclectic collection of recordings, including music by Bach, Schoenberg, Jimi Hendrix, and Led Zeppelin, and then monitored their growth. She reported that the
plants exposed to soft classical music thrived (even when she exposed them to Muzak, that sublime elevator music we all know
and love) while those exposed to *Led Zeppelin II* or Hendrix's
Band of Gypsys were stunted in their growth. To show that it
was in fact the drumbeats of the likes of the legendary drum-
mers John Bonham and Mitch Mitchell that were harming the
plants, Retallack repeated her experiments using recordings of
the same albums but with the percussion blocked out.

As she hypothesized, the plants were not as damaged as they
had been when they were blasted with the full versions, drums
included, of "Whole Lotta Love" and "Machine Gun." Could this
mean that plants have a preferred musical taste that overlapped
with Retallack's? And on a worrisome note, as someone who
grew up studying with Zeppelin and Hendrix blaring from my
stereo system at all times, I wondered when I first encountered
this book if these results implied that I too could have been damaged, as Retallack extrapolates from her results to the effect of
rock music on young people.

Fortunately for me and for the hordes of other Zeppelin fans
out there, Retallack's studies were fraught with scientific short-

comings. For example, each experiment included only a small number of plants (fewer than five). The number of replicates in her studies was so small that it was not sufficient for statistical analysis. The experimental design was poor—some of the studies were carried out in her friend's house—and parameters, such as soil moisture, were determined by touching the soil with a finger. While Retallack cites a number of experts in her book, almost none of them are biologists. They are experts in music, physics, and theology, and quite a few citations are from sources with no scientific credentials. Most important, however, is the fact that her research has not been replicated in a credible lab.

In contrast to Ian Baldwin's initial studies on plant communication and volatile chemicals (encountered in chapter two), which were originally met with resistance by the mainstream science community but subsequently validated in many labs, Retallack's musical plants have been relegated to the garbage bin of science. While her findings were reported in a newspaper article, attempts to publish her results in a reputable scientific journal were unsuccessful, and her book was eventually published as New Age literature. This of course hasn't stopped the book from becoming part of the cultural zeitgeist.

Retallack's results also contradicted an important study published in 1965. Richard Klein and Pamela Edsall, scientists from the New York Botanical Garden, decided to run several tests to determine if plants were truly affected by music. They did this in response to studies coming out of India claiming that music increased the number of branches that would sprout in different plants, one of which was the marigold (*Tagetes erecta*). In an attempt to recapitulate these studies, Klein and Edsall exposed marigolds to Gregorian chant, Mozart's Symphony no. 41 in C major, "Three to Get Ready" by Dave Brubeck, "The Stripper"

Marigold (*Tagetes erecta*)

by the David Rose Orchestra, and the Beatles' songs "I Want to Hold Your Hand" and "I Saw Her Standing There."

Klein and Edsall concluded from their study (which employed strict scientific controls) that music did not influence the growth of the marigolds. As they reported, using humor to convey their general indignation at this line of research, "There was no leaf abscission traceable to the influence of 'The Stripper' nor could we observe any stem nutation in plants exposed to The Beatles."* How can we explain the contradiction between these results and Retallack's subsequent studies? Either Klein and Edsall's marigolds had different musical taste from Retallack's plants, or, more likely, the major methodological and scientific discrepancies in Retallack's study led to unreliable results.

*Nutation is the cyclical swaying or bending movement displayed by different plant parts.

While Klein and Edsall's research was published in a respected professional science journal, it was basically unseen by the general public, and research like Retallack's continued to dominate the popular press in the 1970s. It is also featured prominently in Peter Tompkins and Christopher Bird's iconic book from 1973, *The Secret Life of Plants*, which was marketed as a "fascinating account of the physical, emotional, and spiritual relations between plants and man." In a very lively and beautifully written chapter titled "The Harmonic Life of Plants," the authors reported that not only do plants respond positively to Bach and Mozart but they actually have a marked preference for the Indian sitar music of Ravi Shankar.* Much of the science featured in *The Secret Life of Plants* relied on subjective impressions based on only a small number of test plants. The renowned plant physiologist, professor, and known skeptic Arthur Galston put it succinctly when he wrote in 1974: "The trouble with *The Secret Life of Plants* is that it consists almost exclusively of bizarre claims presented without adequate supporting evidence." But this hasn't kept *The Secret Life of Plants* from influencing modern culture either.

More recent data *supporting* any significant plant response to sound are lacking. However, careful examination of the scientific literature reveals results peppered throughout articles reporting other findings that debunk the idea that plants can hear. In Janet Braam's original paper on the identification of the *TCH* genes (the genes that were activated upon touching a plant), she explained that she checked if in addition to physical stimulation, these genes were induced by exposure to loud music (which for

---

*Some of the shortcomings of Retallack's research are also pointed out in *The Secret Life of Plants*.

her was provided by Talking Heads). Alas, they were not. Similarly, in *Physiology and Behaviour of Plants*, the researcher Peter Scott reported a series of experiments that were set up to test whether corn is influenced by music, specifically Mozart's *Symphonie Concertante* and Meat Loaf's *Bat Out of Hell*. (It's amazing what these types of experiments can tell us about a scientist's own musical taste.) In the first experiment, the seeds exposed to Mozart or Meat Loaf germinated more rapidly than those left in silence. This would be a boon to those who have claimed that music affects plants and a bane to those who think Mozart is qualitatively better than Meat Loaf.

Corn (*Zea mays*)

But this is where the importance of proper experimental controls comes into play. The experiment continued, but this time a small fan circulated any heated air from the speakers away from the seeds. In this new set of experiments, there was no difference in germination rates between the seeds left in silence and those exposed to music. The scientists discovered in the first set of experiments that the speakers playing the music had apparently radiated heat, which improved germination efficiency; the *heat* was the determining factor, not Mozart's or Meat Loaf's music.

Keeping a skeptic's view, let's look again at Retallack's conclusion that the intense drumbeats of rock music are detrimental to plants (and also to people). Could there be an alternative, scientifically valid explanation for loud drumming having a negative effect on plants? Indeed, as I highlighted in the previous chapter, both Janet Braam and Frank Salisbury clearly showed that simply touching a plant a number of times led to dwarfed, stunted plants, or even to a plant's death. So it is conceivable that heavy rock percussion, if pumped through the proper speakers, leads to such powerful sound waves that plants vibrate and are literally "rocked" back and forth as if in a windstorm. In such a scenario we would expect to find reduced growth in plants exposed to Zeppelin, as is the case reported by Retallack. Maybe it's not that the plants don't like rock music; maybe they just don't like being rocked.

Alas, until it is proven otherwise, it looks as if all evidence tells us that plants are indeed "deaf," which is interesting if you consider that plants contain some of the same genes known to cause deafness in humans.

Deaf Genes

The year 2000 was a hallmark for the plant sciences. It was the year that the sequence of the entire *Arabidopsis thaliana* genome was finally communicated to scientists around the world, and they were all eager to hear the results. More than three hundred researchers based at universities and biotechnology companies worked over four years to determine the order of approximately 120 million nucleotides that make up the DNA of the arabidopsis. It also cost approximately seventy million dollars. (The money and the collective effort associated with this project are unfathomable today, as technology has progressed to the point where a single lab can sequence an arabidopsis genome in a little over a week for less than 1 percent of the original cost.)

Arabidopsis was chosen by the National Science Foundation back in 1990 as the first plant that would have its genome sequenced thanks to an evolutionary quirk that resulted in its having relatively little DNA compared to other plants. While arabidopsis has almost the same number of genes (twenty-five thousand) as most plants and animals, it contains very little of a type of DNA called noncoding DNA, which made determining its sequence relatively easy to do. Noncoding DNA is found all over the genome, separating between genes, at the ends of chromosomes, and even within genes. To put things in perspective: while arabidopsis contains about twenty-five thousand genes in 120 million nucleotides, wheat has the same number of genes in 16 *billion* nucleotides (and human beings have about twenty-two thousand genes, fewer than the petite arabidopsis, in 2.9 billion nucleotides).* Because of its small genome, small size, and fast generation time, arabidopsis became the most widely studied plant in the late twentieth century, and as a result, research on this common weed has led to important breakthroughs in many fields. Almost all of the twenty-five thousand genes found in ara-

---

* These numbers should be taken with a grain of salt as the precise definition of "gene" is evolving and with it the numbers. But the general trends and scales are correct.

bidopsis are also present in plants that are important agricultur-
ally and economically, such as cotton and potato. This means that
any gene identified in arabidopsis (say, a gene for resistance to a
particular plant-attacking bacteria) could then be engineered
into a crop to improve its yield.

The sequencing of the arabidopsis and human genomes led
to many surprising findings. Most relevant to our discussion here
is that the arabidopsis genome was found to contain more than a
few genes known to be involved in diseases and disabilities in
humans. (The human genome, on the other hand, contains several genes known to be involved in plant development, such as a
group of genes called the COP9 signalosome that mediate plant
responses to light.) As scientists deciphered the arabidopsis DNA
sequence, they discovered that the genome contains the *BRCA*
genes (which are involved in hereditary breast cancer), *CFTR*
(which is responsible for cystic fibrosis), and a number of genes
involved in hearing impairments.

An important distinction must be made: while genes are often *named* for diseases associated with them, the gene doesn't
exist to *cause* the disease or impairment. A disease occurs when
the gene doesn't function properly as a result of a mutation, which
is a change in the sequence of nucleotides that builds the gene
that disrupts the DNA code. To refresh our understanding of
basic human biology: our DNA code is comprised of only four
different nucleotides, which are abbreviated A, T, C, and G. The
specific combination of these nucleotides provides the code for
different proteins. A mutation or a deletion of a few nucleotides
can catastrophically alter the code. The *BRCA* are genes that,
when mutated or disrupted, can cause breast cancer, but under
normal circumstances they play a key role in determining how
cells know when to divide. When the *BRCA* genes don't function

normally, cells divide too often, and this can lead to cancer. *CFTR* is a gene that, when mutated, when disrupted, causes cystic fibrosis but normally regulates the transport of chloride ions across the cell membrane. When this protein doesn't work properly, chloride ion transport in the lungs (and other organs) is blocked, leading to the accumulation of thick mucus, which manifests clinically as a respiratory ailment.

The names of these genes have nothing to do with their *biological functions*, only their clinical outcome. What are these genes doing in green plants? The arabidopsis genome contains *BRCA*, *CFTR*, and several hundred other genes associated with human disease or impairment because they are essential for basic cellular biology. These important genes had already evolved some 1.5 billion years ago in the single-celled organism that was the common evolutionary ancestor to both plants and animals. Of course, mutations in the arabidopsis versions of these human "disease genes" also disrupt the way the plant functions. For example, mutations in the arabidopsis breast cancer genes lead to a plant whose stem cells (yes, arabidopsis has stem cells) divide more than normal cells, and the whole plant is hypersensitive to radiation, both of which are also hallmarks of human cancer.

This puts in perspective what a "deaf" gene is: a gene that—when mutated—leads to deafness in humans. More than fifty human "deaf" genes have been identified by various labs worldwide, and at least ten of these fifty also show up in arabidopsis. Just because deaf genes were discovered in the arabidopsis genome doesn't mean that the plant can hear, just as the presence of *BRCA* in arabidopsis doesn't mean plants have breasts. The human "deaf" genes have a cellular function necessary for the ear to work properly, and when any of these genes contain a mutation, the result is hearing loss.

Four of the arabidopsis genes connected to hearing impairment encode very similar proteins called myosins. Myosins are known as motor proteins, because they work as "nanomotors" that literally carry and move different proteins and organelles around the cell.* One of the myosins involved in hearing helps form the hair cells in the inner ear. When this myosin contains a mutation, our hair cells don't form properly, and they don't respond to sound waves. In the plant world, we find that plants have hairlike appendages on their roots, aptly referred to as root hairs which help roots soak up water and minerals from the soil. When a mutation occurs in one of the four arabidopsis "deaf" myosin genes, the root hairs don't elongate properly, and consequently the plants are less efficient at absorbing water from the soil.

Myosin and the other genes found in both plants and humans have similar functions at the cellular level. But when you put all the cells together, the function for the particular organism is different: we need myosin to facilitate the proper functioning of our inner-ear hairs and, ultimately, to hear; plants need myosin for the proper functioning of their root hairs, which allows them to drink water and find nutrients from the soil.

The Deaf Plant

Serious and reputable scientific studies have concluded that the sounds of music are truly irrelevant to a plant. But are there sounds that, at least theoretically, could be advantageous for a plant to respond to? Professor Stefano Mancuso, director of the

---

*This website illustrates myosin in action: www.sci.sdsu.edu/movies/actin_myosin_gif .html.

International Laboratory of Plant Neurobiology at the University of Florence, has recently been using sound waves to increase yield in a vineyard in the Tuscany wine region. But the basic biology behind this agricultural use of sound waves is still unclear.

Dr. Lilach Hadany, a theoretical biologist at Tel Aviv University, uses mathematical models to study evolution. She proposes that plants do respond to sounds but that we have yet to carry out the correct experiments to detect their doing so. Indeed in science in general, lack of experimental evidence does not equate to a negative conclusion. In her mind, we would have to construct a study in which we would use a sound from the natural world known to influence a specific plant process. One such sound could be the buzzing of bees. In a process referred to as buzz pollination, bumblebees stimulate a flower to release its pollen by rapidly vibrating their wing muscles without actually flapping their wings, leading to a high-frequency vibration. While this vibration can be heard (it's the buzz we hear when a bee flies by), the pollen release necessitates a physical contact between the vibrating bee and the flower. So just as deaf people can feel and respond to vibrations in music, flowers feel and respond to bumblebee vibrations, without necessarily *hearing* them. But conceivably, the sound of the vibrations could also affect the flower in some yet undetected way.

In a similar vein, Roman Zweifel and Fabienne Zeugin from the University of Bern in Switzerland have reported ultrasonic vibrations emanating from pine and oak trees during a drought. These vibrations result from changes in the water content of the water-transporting xylem vessels. While these sounds are passive results of physical forces (in the same way that a rock crashing off a cliff makes a noise), perhaps these ultrasonic vibrations are used as a signal by other trees to prepare for dry conditions?

If scientists are going to properly study plant responses to sound waves, we need to understand that *if* a plant needs to hear, then its auditory system would be much different from what has evolved in animals. In addition to the few examples above, perhaps some plants sense minuscule sounds that might be created by tiny organisms. That kind of system may be off the radar screen of most physiological tools.

While these possibilities are interesting to ponder, in lieu of any hard data to the contrary we must conclude for now that plants are deaf and that they did not acquire this sense during evolution. The great evolutionary biologist Theodosius Dobzhansky wrote, "Nothing in biology makes sense except in the light of evolution." Considering this, perhaps we can understand why hearing, as opposed to the other senses we've covered, isn't really needed by plants.

The evolutionary advantage created from hearing in humans and other animals serves as one way our bodies warn us of potentially dangerous situations. Our early human ancestors could hear a dangerous predator stalking them through the forest. We notice the faint footsteps of someone following us late at night on a poorly lit street. We hear the motor of an approaching car. Hearing also enables rapid communication between individuals and between animals. Elephants can find each other across vast distances by vocalizing subsonic waves that rumble around objects and travel for miles. A dolphin pod can find a dolphin pup lost in the ocean through its distress chirps, and emperor penguins use distinct calls to find their mates. What's common in all of these situations is that sound enables a rapid communication of information and a response, which is often movement—fleeing from a fire, escaping from attack, finding family.

As we've seen, plants are sessile organisms, secured to the ground by their roots. While they can grow toward the sun and bend with gravity, they can't flee. They can't escape. They don't migrate with the seasons. They remain anchored in the face of an ever-changing environment. Plants also operate on a different timescale from animals. Their movements, with the conspicuous exception of plants like *Mimosa* and the Venus flytrap, are quite slow and are not easily noted by the human eye. As such, plants have no need for detailed communication that could allow for a quick retreat. The rapid audible signals we're used to in our world are irrelevant to a plant. Plants lack the structures for purposeful vocalization, and the sounds of leaves in the wind or branches cracking under our feet do not communicate anything to the plant. For hundreds of millions of years plants have thrived on earth, and the nearly 400,000 species of plants have conquered every habitat without ever hearing a sound. But although they may be deaf, plants are acutely aware of where they are, what direction they're growing, and how they move.

FIVE

How a Plant Knows Where It Is

I never saw a discontented tree. They grip the ground as
though they liked it, and though fast rooted they travel
about as far as we do. They go wandering forth in all
directions with every wind, going and coming like ourselves, traveling with us around the sun two million
miles a day, and through space heaven knows how fast
and far!

—John Muir

Shoots grow up; roots grow down. That seems simple enough,
but how do plants know where up is? You might think it's all due
to sunlight, but if light is a plant's main signal for up, how would
it know where up is at night? Or when it's merely a seed germi-
nating underneath the soil? And you might think that down has
to do with touching dark moist soil. But the aerial roots of the
banyan and mangrove trees also always grow down, even though
they start several meters up in the air.

Scientists have documented that when a plant has been
turned upside down, it will reorient itself in a slow-motion
maneuver—like when a cat is falling and rights itself before it

lands—so that its roots grow down and its shoots grow up.* And not only do plants know when they're upside down, but experiments have also proven that they're constantly aware of where their branches are; they know if they're growing perpendicular to the ground or at an angle off to one side, and tendrils always have a pretty good idea of where the nearest support is to grab onto. Just think of the dodder plant that makes *circles* in the air while it searches for a suitable plant to parasitize. But how does a plant really know where it is in space? How is it that *we* know?

We know because of our sixth sense, and contrary to popular belief the sixth sense is not ESP; it's proprioception. Proprioception enables us to know where different body parts are relative to each other, without having to look at them. While our other senses are *outwardly* oriented, receiving signals like light, odor, and sound from external sources, proprioception provides us with information based solely on the internal status of the body. It enables you to move your legs in a coordinated fashion to walk, to swing your arm to catch a baseball, and to scratch an itch on the back of your neck. Without proprioception, a simple task like brushing your teeth would be practically impossible.

It's the kind of sense that we don't much consider unless we lose it. If you've ever been even slightly inebriated, you've experienced compromised proprioception. It's why police officers use a field sobriety test on drivers suspected of intoxication; the test, which involves simple physical "hand-eye coordination" tasks, easily reveals who has impaired proprioception and who does not. When you're sober, touching your nose with your eyes closed

---

*This clip, http://phytomorph.wisc.edu/assets/movies/gravitropism.swf, shows a time-lapse movie of a root that's placed on its side and slowly but surely turns to grow down. Other great movies can be found at http://plantsinmotion.bio.indiana.edu.

is a simple task. But people who are even moderately drunk will find this easy test much more difficult.

Understanding proprioception is less intuitive than understanding our other senses because it lacks a clear focus organ. Vision is perceived through the eyes; olfaction through the nose; and hearing through the ears. Even tactile sensation through the nerves in the skin is easy enough to grasp. Proprioception, on the other hand, involves the coordinated input of signals from the inner ear, which communicates balance, along with signals from specific nerves throughout the body that communicate position.

Adjacent to the inner-ear structures necessary for hearing is a complex system of miniature chambers called the semicircular canals and the vestibule, which work together to sense the position of your head. The semicircular canals lie at right angles to each other, forming a structure that resembles a gyroscope. The canals are filled with fluid, and when our head changes position, the fluid moves. Sensory nerves at the base of each canal respond to the waves in the fluid, and because the canals are situated in three distinct planes, they're able to communicate movement in all directions. The vestibule is also filled with fluid, and it contains sensory hairs as well as otoliths, small crystalline stones that literally sink in response to gravity, adding extra pressure (and thus stimulation) on the sensory hairs in the vestibule. This informs us if we are upright, horizontal, or upside down. The pressure of the otoliths on the nerves in different areas of the vestibule helps us differentiate up from down. This function is thrown out of whack on some amusement park rides that shake the otoliths around so much, we have no idea where we are.

While the workings of the inner ear help us keep our balance, the proprioceptive nerves throughout our body keep everything coordinated, and the proprioceptive receptors inform the brain of the position of our limbs. These nerves are distinct from tactile nerves that sense pressure or pain and are located deep within our bodies in our muscles, ligaments, and tendons. The anterior cruciate ligament in the knee, for example (also known as ACL), contains nerves that communicate proprioceptive input from the lower leg. A few years ago, I tore my ACL after being challenged by my son on the ski slopes. I was very surprised to discover after the accident that I had trouble walking: I kept tripping over my own foot. I had lost the proprioceptive positioning signaling of my foot—and eventually regained it as my brain began to reintegrate information from other nerves in my lower leg.

Two main interrelated bodily processes depend on proprioception—being aware of the relative position of parts of the body while at rest (static awareness), and being aware of the relative position of parts of the body while in motion (dynamic awareness). Proprioception encompasses not only our sense of balance but our coordinated motion as well—from the simple wave of a hand, to the more complicated integration of movement and balance you need to walk down the street, to the very complex movements of an Olympic gymnast performing a somersault on a balance beam. These two processes—static and dynamic awareness of body position—are also interrelated in plants and have been a focus of many botanists for years.

Knowing Up from Down

In 1758—more than a century before Darwin's landmark book *The Power of Movement in Plants*—Henri-Louis Duhamel du Monceau, a French naval inspector and ardent botanist, observed that if he turned a seedling upside down, its root would reorient itself in order to grow down, while its shoot would bend and grow up toward the sky. This simple observation of roots growing as if pulled down by gravity (positive gravitropism) and shoots growing in the opposite direction against this pull (negative gravitropism) led to a number of questions and hypotheses that have continued to influence research carried out in labs around the world. Many scientists who read what Duhamel had to say concluded that the ways in which the roots reoriented themselves had to do with gravity indeed. But Thomas Andrew Knight, a fellow of the Royal Society, pointed out about fifty years later that "the hypothesis [of gravity affecting plant growth] does not appear to have been strengthened by any facts." While many scientists interpreted Duhamel's observation as proof that gravity influences the ways a plant grows, none had carried out rigorous scientific experimentation to test this idea, which is what Knight set out to do.

Knight was part of the landed gentry and lived in a castle in the West Midlands region of England, surrounded by extensive gardens, orchards, and greenhouses. He was not trained as a scientist, but as was common for nineteenth-century aristocrats, he used his leisure time to pursue scientific knowledge, and he soon became especially skilled in horticulture. In fact, he turned out to be one of the leading plant physiologists of his time. For his studies on how plants know up from down, Knight developed a very sophisticated experimental apparatus that negated the effect of Earth's gravity on plant growth while simultaneously applying a new centrifugal force that would act on the roots. He constructed a waterwheel that was turned by a stream running through his estate, and he attached a wooden plate to the wheel so that the plate turned with the wheel. He fastened several bean seedlings around the plate in various positions so that their root tips were facing in all possible directions—into the center, out, at an angle, and so on.

He let the wheel spin at the nauseating speed of 150 revolutions per minute for several days. The seedlings somersaulted with each spin of the plate. At the end of the treatment, Knight saw that all of the roots had grown *out* from the center of the wheel, while all the shoots grew toward the center.

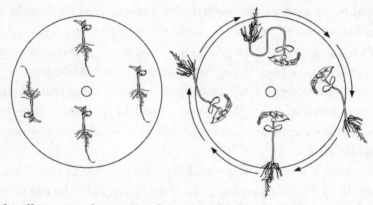

This illustration depicts Knight's waterwheel with seedlings on it before the experiment started and at its end.

With his makeshift centrifuge, Knight had applied a force on the seedlings that mimicked gravitation and demonstrated that the roots always grew in the direction of this centrifugal force—while the shoots grew in the opposite direction. Knight's work provided the first experimental corroboration for Duhamel's observations. He showed that roots and shoots respond not only to natural gravity, as Duhamel showed, but also to an artificial gravitational force supplied by his waterwheel-powered centrifuge. But this still didn't explain *how* a plant could sense gravity.

Interest in how plants sense gravity picked up again toward the end of the nineteenth century. As with so many questions in the plant sciences, it was Darwin and his son Francis who performed the definitive experiments in the field, and in true Darwinian fashion they carried out an extremely detailed, exhaustive study, in this case to determine precisely which part of the plant senses gravity. Their initial hypothesis was that "gravireceptors" (analogous to photoreceptors for light) were located in the tip of the root. To test their hypothesis, they sliced off different lengths of the root tips of beans, peas, and cucumbers and then placed the roots on their sides over damp soil. While the roots continued to elongate, they no longer had the ability to reorient their growth and bend down into the soil. Even the amputation of only 0.5 mm of the tip was enough to obliterate the plant's overall sensitivity to gravity! The Darwins also noticed that if the tip of a root grew back within several days of the amputation, the root would regain its ability to respond to gravity and would go back to its old ways, bending down into the soil.

This result was similar to what Darwin discovered when he conducted his research in phototropism. In his phototropism experiment, he showed that the tip of a *shoot* sees the light and transfers this information to its midsection to tell it to bend toward the light. Here, Darwin and his son showed that the tip of the root feels the gravity, even though the bending occurs farther up the root. From this Darwin further hypothesized that the root tip somehow sent a signal up the rest of the root to tell it to grow down with the gravity vector.

To test this hypothesis, Darwin placed a bean seedling on its side and immobilized it with a pin on the top of some soil, but this time he waited ninety minutes before he amputated the root tip (with a normal plant placed on its side, it usually takes several hours before the root's reorientation is obvious). He found that the root still reoriented downward even though it was "tipless." During the ninety minutes before he severed the tip, Darwin assumed the bean plant had sent instructions up the root that told the plant to bend down. Darwin and his son witnessed the same results in similar experiments with six different types of plants and in cases in which they burned the tip with silver nitrate instead of amputating it. They concluded that the root tip must immediately sense gravity and then pass this information along, telling the plant which direction is optimal for its growth.

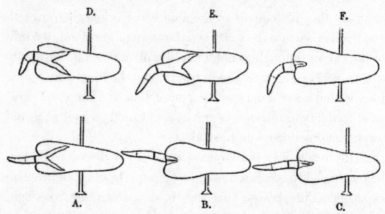

Darwin secured bean (*Vicia faba*) seedlings on their sides with pins for twenty-three hours and thirty minutes. In A, B, and C, he cauterized the root tips with silver nitrate (rather than simply amputating the root tip). The roots in D, E, and F were left untreated.

Our understanding of how a plant knows up from down progressed remarkably over the course of the eighteenth and nineteenth centuries. First Duhamel revealed that seedlings reorient their growth so that roots grow down and shoots up, then Knight demonstrated that gravity was the reason for this "up-down growth," and then the Darwins showed that the root tip contains the mechanism that senses the gravity. More than a century would pass before modern molecular genetic studies would confirm Darwin's results, demonstrating that the cells in the extreme end of the root (in a region called the root cap) sense gravity and help a plant know where down is.

If a plant needs its root tips to be intact in order to grow down to the ground, you might expect (as Darwin did) that the tip of the shoot is essential for the plant to grow up toward the sky. After all, Darwin showed that cutting off the top part of a plant causes it to lose its ability to see and bend to light coming from the side. But surprisingly, it turns out that a plant that's had the tip of its shoot chopped off will still grow up; it retains its ability for negative gravitropism. Could this mean that the root and the shoot sense gravity in different ways?

Much of our current understanding of the ways in which plants perceive gravity has come from studies that use everyone's favorite laboratory plant, arabidopsis. Just as Maarten Koornneef and his colleagues isolated "blind" plants that were defective in different photoreceptors (as we saw in chapter one), many scientists have isolated mutant arabidopsis plants that don't know up from down. The procedure is actually quite simple: scientists grow thousands of mutant arabidopsis seedlings for a week and then flip their containers by ninety degrees. Almost all the seedlings will reorient so that the shoot grows up and the roots down. But the rare mutant that doesn't sense the gravity will keep growing without any change in its direction.*

---

*In these types of studies, the seeds are often first treated with a chemical that causes mutations in the DNA. The chance of the chemical working on a specific gene needed for gravitropism is very small, so thousands of seedlings have to be tested. Luckily, arabidopsis seedlings are minuscule, so screening through such a large number is possible.

Many of these mutants have defects in their roots as well as their shoots, and they've lost the ability to know up from down. But in other mutant arabidopsis, only the root or the shoot is affected, which suggests that they detect gravity in different ways. For example, an arabidopsis mutant in the gene termed *scarecrow* has shoots that don't know when they've been placed on their side, so the mutant plant stays horizontal (it's defective in negative gravitropism in the shoot).* But surprisingly, roots in this mutant know how to grow down (the roots maintain positive gravitropism). A Japanese morning glory cultivar called *Shidare-asagao* (which means "weeping") has shoots that don't know up from down; sure, it makes for an attractive ornamental hanging plant, but it also provides scientists with a great mutant to study gravitropism. What makes this plant's stems and leaves grow in varied directions? Recent genetic studies demonstrate that *Shidare-asagao* contains a mutation in its *scarecrow* gene, in fact. Which begs the question: Do these mutants ultimately prove that the mechanism for sensing gravity differs in the parts of the plant above- and belowground?

Actually, this mutant doesn't tell us that the *mechanism* for sensing gravity is different in the root and stems, but it does tell us that the specific *site* is different (which we already knew from Darwin's studies). Scientists in Phil Benfey's lab at New York University used the *scarecrow* mutant to figure out which part of the stem senses gravity. Around the turn of the twenty-first cen-

---

*Naming mutants in arabidopsis, and indeed in other organisms, is the prerogative of the scientist who first isolates the mutant. The name of the mutant is presented in lowercase letters in italics and corresponds to the name of the mutant gene. Some scientists are more conservative and name their mutants after their obvious characteristics (such as the *shortroot* mutant of arabidopsis, which has, no surprise, short roots). Others are more creative. Examples of names of arabidopsis mutants include *scarecrow*, *toomanymouths*, and *werewolf*.

Morning glory (*Pharbitis nil*)

tury they discovered that the *scarecrow* gene is necessary for the formation of the endodermis, a group of cells that wrap around the vascular tissues of the plant. In the roots, the endodermis acts as a selective barrier that actively regulates how much and which compounds (such as water, minerals, and ions) enter the xylem tubes for transport to the green parts of the plant. Plants that have a mutated *scarecrow* gene don't have any endodermis. While this makes for rather short and weak roots, they still know how to grow down. They know this because the root's gravisensors in the tip do not contain endodermis cells. The *scarecrow* mutant still has a normal root tip, so it knows where down is.

But if shoots don't have an endodermis, they can't know where up is, and that's as detrimental to a plant's sense of direction as amputating the root tip. In other words, two distinct plant tissues detect gravity in the lower and aerial parts of the plant. In the roots, it's the root tip; in the stem, it's the endodermis. So while our "gravireceptors" are only in the inner ear, plants have them in many places in their root tips and in their stems.

How do these specific groups of plant cells in the root tip and in the endodermis sense gravity? The first answers came from studies of the root cap using a microscope to see their incredible subcellular structures. Cells in the central area of the root cap contain dense ball-like structures called statoliths (derived from the Greek for "stationary stone"), which—similar to the otoliths in our ears—are heavier than other parts of the cell and fall to the bottom side of the cells of the root cap.* When a root is placed on its side, the statoliths fall to the new bottom of the cell just as marbles would roll to the bottom part of a jar on its side. Not surprisingly, the only aerial plant tissue that contains statoliths is the endodermis. Just as in the root cap, when a plant is on its side, the statoliths in the endodermis fall onto what was the side of the cell, and this part becomes the new bottom of the plant. The ways in which the statoliths responded to gravity led scientists to propose that they are in fact the gravity receptors.

If statoliths are the plant gravity receptors, then the simple displacement of statoliths should be sufficient to cause a plant to change its direction of growth as if it had been affected by gravity. Only with the advent of molecular genetics and, interestingly enough, spaceflight—which I'll get to in a moment—have scientists been able to carry out the experiments that engage with this question.

For the past twenty years, John Kiss and his colleagues at

---

*Statoliths in higher (flowering) plants are also known as amyloplasts, modified forms of chloroplasts that contain starch rather than chlorophyll.

Miami University in Ohio have been using some of the coolest toys in science to determine whether statoliths really are what sense gravity in a plant. Using a high-gradient magnetic field that simulates gravity, Kiss induced his statoliths to migrate laterally as if he had turned the plants on their sides. When this happens, the roots start to bend in the same direction that the statoliths move: if the statoliths move to the right, the root bends to the right; if the statoliths move to the left, the root bends to the left. These results really supported the idea that the position of statoliths is what tells a plant where down is. They also led Kiss to predict that in the absence of gravity, statoliths wouldn't fall to the bottom of a cell, and thus a plant won't know where down is. Of course to test such a hypothesis, Kiss would need conditions with no gravity, as in a spacecraft orbiting Earth.

Aboard the space shuttle, where plants obviously don't experience the effects of gravity, the statoliths can't fall, and they remain naturally distributed throughout the cell. Under these weightless conditions, Kiss could not detect any gravitropic bending in the plants that were out in space. These studies revealed a fascinating clue to why plants move the way they do: a plant needs statoliths to sense gravity, just as we need otoliths in our ears to stimulate our balance receptors.

The Movement Hormone

The ways in which an inverted bean root responds to gravity, and a tulip in a window box moves toward the sun, and a *Cuscuta* sidles up to the neighboring tomato are similar: the plants sense a change in their environments (gravity, light, or smell) and bend in response to the stimulus. The stimuli are diverse, but the responses are similar—growth in a particular direction. We've dealt a lot with how a plant senses gravity (and light and smells), but we haven't explored how this sensory information tells a plant to grow and bend. Let's reexamine Darwin's experiments on phototropism from the first chapter. He showed that the tip of the grass seedling "sees" the light and transfers this information to its midsection in order to tell it to bend toward the light. This is similar to the root cap "feeling" the gravity and then transferring the information up the root to induce the plant to grow downward, or to the *Cuscuta* that smells the tomato and then bends toward it.

At the beginning of the twentieth century, the Danish plant physiologist Peter Boysen-Jensen expanded on the Darwins' experiments on phototropism. Like Darwin, he cut off the tips of his oat seedlings, but before returning the tips to their plant stumps, he did something unusual but quite brilliant in its way. He placed either a thin slab of gelatin or a tiny piece of glass between the stump and the tip. When he illuminated these plants from the side, the one with the gelatin slice bent toward the light, while the one with the glass stayed straight. This proved to Boysen-Jensen that the bending signal coming from the tip of the plant must be soluble since it could clearly pass through the gelatin but not the glass. Yet Boysen-Jensen didn't know what the chemical was that traveled from the tip down to the stalk to make it bend.

In the early 1930s scientists finally identified the growth-promoting chemical that passed from the tip through the gelatin and down the stock and called it auxin, which derives from the Greek for "to increase." While plants have many different hormones, none are as prevalent or involved in as many processes and functions as auxin. One of these functions is to tell cells to increase their length. Light causes auxin to accumulate on the dark side, causing the stem to elongate on the dark side only,

Oats (*Avena sativa*)

which results in the stem bending toward the light. Gravity makes the auxin appear on the "up side" of roots, which causes them to grow down, and on the "down side" of stems and leaves, causing them to grow up. While different stimulations activate different plant senses, many of the plant's sensory systems converge on auxin, the movement hormone.

Dancing Plants

As mentioned earlier in this chapter, proprioception is more than just knowing up from down; it's also knowing where the body parts are when you are moving. When Mikhail Baryshnikov leaps across a stage and lands in an arabesque, he's not only perfectly balanced but also acutely aware of the position of every part of his body. He knows how far his leg is stretched out behind him, how high his hand is from his shoulder, the exact tilt of his torso.

Of course, it's no surprise that we regard plants as stationary beings; they are sessile organisms that are eternally rooted and incapable of locomotion. But when we observe them patiently over a long period of time, this stationary stature gives way to an intricately choreographed festival of movement, much like Baryshnikov springing to life in the first scene of a ballet. Leaves curl and unfold, flowers open and close, and stems circle and bend.

These movements are best seen in time-lapse photography, and one of the first uses of time-lapse photography was, in fact, to do just this. Professor Wilhelm Pfeffer, who trained with Darwin's friend Julius von Sachs, filmed a variety of plants in motion, from tulips to *Mimosa* to broad beans. His early movies are grainy yet fascinating to watch.° Long before time-lapse photography came into play, however, the persistent and dogged Darwin studied plant movements using a very time-consuming, low-tech procedure: he suspended a glass plate above a plant and marked on the glass the position of the tip of the plant every few minutes for several hours. By connecting the dots, he mapped out the exact movements of his subject. (An insomniac, Darwin undoubtedly spent more than a few nights meticulously monitoring the more than three hundred different species that he eventually recorded in this fashion, including the wild cabbage depicted on the following page.)

Darwin found that all plants move in a recurring spiral oscillation, which he termed "circumnutation" (Latin for "circle" or "sway").† This spiral pattern varies between species and can range from a repeating circle to an ellipse to a trajectory of inter-

---

°For examples, see www.dailymotion.com/video/x1hp9q_wilhem-pfeffer-plant-movement
_shortfilms#from=embed.
†For a good example of circumnutation, look at this movie: www.pnas.org/content
/suppl/2006/01/11/0510471102.DC1/10471Movie1.mov.

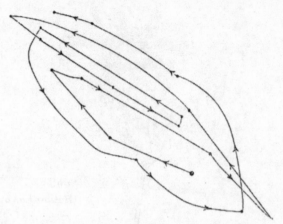

Darwin's trace of the movements of the tip of a wild cabbage (*Brassica oleracea*) seedling over ten hours and forty-five minutes

locking shapes much like the images from a Spirograph. Some plants have surprisingly large movements, such as bean shoots, which circle in a radius of up to ten centimeters. Others move in millimeters, like strawberry branches. Speed is another variable; tulips circumnutate at a rather fixed speed (they take about four hours), while other plants vary significantly: arabidopsis stems take between fifteen minutes and twenty-four *hours* for one circle, and wheat usually completes a rotation once every two hours. We don't know what the basis for this individuality in movement is, but we do know that both environmental and internal factors can influence speed. As the Polish scientist Maria Stolarz found, if she used a small flame to burn a sunflower leaf for only three seconds, the circling time of the plant almost doubled for one rotation. Then the sunflower would bounce back to its initial rate.

Darwin was fascinated by these movements, and he came to the conclusion not only that circumnutation was hardwired into

Sunflower
(*Helianthus annuus*)

the behavior of all plants but that these spiraling oscillatory dances were actually the driving force for all plant movements. He proposed that phototropism and gravitropism were just modified circumnutations aimed in a specific direction. This hypothesis remained unchallenged until some eighty years later, when Donald Israelsson and Anders Johnsson at the Lund Institute of Technology came up with an alternative hypothesis that the oscillatory movements of plants were simply a *result* of gravitropism (and not the *cause*). As a plant grows, they argued, a slight change in the position of the stem (whether caused by wind, light, or a physical barrier) will result in the displacement of the statoliths, which in turn will trigger the stem to bend up, even as external factors push its position around a bit.

This bending, however, often overshoots its goal. Just like those old Bozo the Clown punching bags that keep bouncing back up at you, when a stem reorients itself vertically, it first overshoots straight up and down and bends somewhat in the opposite direction. Now that the stem isn't straight again, but oriented in the other direction, the statoliths redistribute a second time, initiating a gravitropic response toward the opposite side of the plant. This new growth will also overshoot, and the cycle repeats, leading to the classic oscillatory movement that Darwin documented for cabbage and clover and that we see in tulips and cucumbers. Just as Bozo the Clown goes back and forth in circles trying to find its center, the stem of the plant circles in the air as it searches for balance.

So Darwin hypothesized that these dances are a built-in behavior of all plants, while Israelsson and Johnsson believed that gravity powers the circling dances of plants. And finally, the two competing theories could be put to the test by the end of the twentieth century with the advent of spaceflight. If Darwin's theory is correct, circumnutations would continue unimpeded in the absence of gravity; if Israelsson and Johnsson's statolith-centered model is correct, the circumnutation in plants would not occur in space.

Back in the infancy of the space program in the 1960s, Allan H. Brown, a well-known and respected plant physiologist, conceived one of the first experiments with arabidopsis in space in what was part of the Biosatellite III program. Brown wanted to test whether plant movements would continue in the absence of gravity.° When the program was canceled due to budget cuts, Brown had to wait until 1983, when his experiments on plants were among the first to be carried out on the space shuttle. The astronauts on board the space shuttle *Columbia* monitored the movements of sunflower seedlings while they were in orbit and transmitted the data to scientists on Earth. Sunflower seedlings exhibit robust movements on Earth, so they were the ideal plant

°Actually, referring to orbit we say "microgravity" rather than "no gravity" as there is still a small gravitational pull, of about 0.001 percent.

to lob up with the shuttle to see what would happen in space. Aboard *Columbia*, miles and miles above Earth, almost 100 percent of the seedlings exhibited rotational growth patterns; even in the near absence of gravity, the sunflower seedlings continued their gyrations in the same way they did on Earth. This strongly supported Darwin's theory.

But let's revisit the second hypothesis: that spiraling is intimately connected to gravity. A few years ago, Hideyuki Takahashi and his colleagues at the Japan Aerospace Exploration Agency monitored circumnutation in the morning glory mutant that lacked a gravity-sensing endodermis in its shoot. The morning glory mutant that didn't respond to gravity also didn't move in the spiraling style that a normal morning glory does. Furthermore, arabidopsis mutants that have small or defective statoliths also didn't spiral. These results wouldn't have made Darwin happy: they strongly support the idea that circumnutation and gravitropism are intimately connected (of course, Darwin would likely have appreciated the science here, modified his own hypotheses, and developed new experiments to test them).

Takahashi explained the contradiction between his results and those garnered on *Columbia* by suggesting that since the experiments on the shuttle were carried out on seeds *germinated* on Earth, this may have been enough to perpetuate circumnutation out in space. Indeed, it would make sense that a seed formed on Earth would have different characteristics from one formed in space, and if so, the time limitation of the experiments carried out on *Columbia* (about ten days) may have influenced the outcome of the experiment.

The International Space Station, which began operation in 2000, finally provided a facility for long-term experiments on the effect of gravity on plants. Anders Johnsson could put his nearly forty-year-old hypothesis to the test when he and his Norwegian colleagues carried out an important experiment aboard the space station over several months in 2007. Their setup consisted of arabidopsis plants that germinated aboard the space station and were grown in a special chamber designed for use in space. They were automatically photographed every few minutes in order to monitor their exact positions and detect any movements. In the near-weightless conditions of the space station, the arabidopsis plants exhibited spiral patterns of movement, albeit very minute ones, demonstrating the movements that Darwin had predicted and confirming Brown's own observations. But the radius of the circular movement, and the speed of movement, were less than those found on Earth, suggesting that gravity was essential for amplifying the built-in movement.

These weightless plants were placed on a large rotating centrifuge that mimicked gravitation, much as Knight's waterwheel had years and years earlier. The plants could be continuously monitored by a camera while they rotated. Very soon after feeling the g-force, the plants started moving in more exaggerated circles. Both the size and the speed of the movements of the spinning plants were similar to those detected on the arabidopsis plants grown on Earth. This revealed that gravity is not necessary for the movements, but rather modulates and amplifies the endogenous movements in the plant. Darwin was correct: as far as we know, circumnutation is a built-in behavior of plants, but this behavior needs gravity to reach its full expression.*

*The overall mechanism of sensing gravity is more complex than simply statoliths falling within the cell.

The Balanced Plant

A plant can be pulled in many directions at once. Sunlight hitting a plant at an angle causes it to bend toward the rays, while the sinking statoliths within the plant's bending branches tell it to straighten up. These often conflicting signals enable a plant to situate itself in a position that is optimal for its environment. The tendrils of a vine, searching for a support to grab onto, will be attracted to the shade of the neighboring fence, and gravity will enable the vine to rapidly twirl around it. A plant on a windowsill will be pulled by the light and grow to one side, toward the sunny part of the sill, while the force of gravity will influence it to grow up at the same time. The smell of the tomato will pull *Cuscuta* to the side, while gravitropism will push it to keep growing upward. Just like Newtonian physics, the position of any part of the plant can be described as a sum of the force vectors acting upon it that tell a plant both where it is and which direction to grow.

Humans and plants respond to gravity in similar ways and rely on our sensors to inform us of our positions and balance. But when we move, we not only know where our body parts are in relation to others but also remember the movement, which allows us to repeat the movement again and again. Can a plant remember its past movements too?

SIX

What a Plant Remembers

The oaks and the pines, and their brethren of the wood,
have seen so many suns rise and set, so many seasons
come and go, and so many generations pass into silence, that we may well wonder what "the story of the
trees" would be to us if they had tongues to tell it, or we
ears fine enough to understand.

—Maud van Buren, *Quotations for Special Occasions*

Memories often take up a good portion of an average person's
daily mental wanderings. We may remember an especially savory
feast, the games we played as children, or a particularly humorous incident at the office from the day before. We can envision a
breathtaking sunset that we once saw on the beach, and we also
remember significantly traumatic and scary experiences. Our
memory is dependent on sensory input: a familiar smell or a favorite song can trigger a wave of detailed memory that transports
us back to a particular time and place.

As we've seen, plants benefit from rich and varied sensory
inputs as well. But plants obviously don't have memories in the
way we do: they don't cower at the thought of a drought or dream

about the sunbeams of summer. They don't miss being encased inside a seedpod, nor do they feel anxious about premature pollen release. Unlike Grandmother Willow in Disney's *Pocahontas*, old trees don't remember the history of the people who have slept in their shade. But as we've seen in earlier chapters, plants clearly have the ability to retain past events and to recall this information at a later period for integration into their developmental framework: Tobacco plants know the color of the last light they saw. Willow trees know if their neighbors have been attacked by caterpillars. These examples, and many more, illustrate a delayed response to a previous occurrence, which is a key component to memory.

Mark Jaffe, the same scientist who coined the term "thigmomorphogenesis," published one of the first reports of plant memory in 1977, though he didn't call it as such (instead, he talked about one- to two-hour retention of the absorbed sensory information). Jaffe wanted to know what makes pea tendrils curl when they touch an object suitable to wrap themselves around. Pea tendrils are stem-like structures that grow in a straight line until they happen upon a fence or a pole they can use for support, and then they rapidly coil around the object to grab onto it.

Jaffe demonstrated that if he cut a tendril off of a pea plant but kept the excised tendril in a well-lit, moist environment, he could get it to coil simply by rubbing the bottom side of the tendril with his finger. But when he conducted the same experiment in the dark, the excised tendrils didn't coil when he touched them, which indicated that the tendrils needed light to perform their magic twirling. But here was the interesting catch: if a tendril touched in the dark was placed in the light an hour or two later, it spontaneously coiled without Jaffe having to rub it again.

Somehow, he realized, the tendril that had been touched in the dark had stored this information and recalled it once he placed it in the light. Should this storage and later recollection of information be considered "memory"?

In fact, research on human memory conducted by the renowned psychologist Endel Tulving provides us with an initial foundation from which to explore plants and their unique "recollections." Tulving proposed that human memory exists on three levels. The lowest level, procedural memory, refers to nonverbal remembering of *how* to do things and is dependent on the ability to sense external stimulation (like remembering to swim when you jump in a pool). On the second level is semantic memory, the memory of concepts (like most of the subjects we learned in school). And the third level is episodic memory, which refers to remembering autobiographical events, like funny costumes from childhood Halloween parties or the loss we felt at the death of a dear pet. Episodic memory is dependent on the "self-awareness" of the individual. Plants clearly do not make the cut for semantic and episodic memory: these are the memories that define us as human beings. But plants are capable of sensing and reacting to external stimulation, so by Tulving's definition plants should be capable of procedural memory. And indeed, Jaffe's pea plants illustrate this. They sensed Jaffe's touch, remembered it, and coiled in response.

Neurobiologists know quite a bit about the physiology of memories and can pinpoint the distinct (but still interconnected) areas of the brain that are responsible for different types of memory. Scientists know that electric signaling between neurons is essential for memory formation and storage. But we know much less about the molecular and cellular basis of memory.

What's fascinating is that the latest research hints that while memories are infinite, only a very small number of proteins are involved in memory maintenance.

We need to be aware, of course, that what we refer to as "memory" for people is actually a term that encompasses many distinct forms of memory, beyond the ones described by Tulving. We have *sensory memory*, which receives and filters rapid input from the senses (in a blink of an eye); *short-term memory*, which can hold up to about seven objects in our consciousness for several seconds; and *long-term memory*, which refers to our ability to store memories for as long as a lifetime. We also have *muscle-motor memory*, a type of procedural memory that is an unconscious process of learning movements such as moving fingers to tie a shoelace; and *immune memory*, which is when our immune systems remember past infections in order to avoid future ones. All but the last are dependent on brain functions. Immune memory is dependent on the workings of our white blood cells and antibodies.

What's common to all forms of memory is that they include the processes of forming the memory (encoding information), retaining the memory (information storage), and recalling the memory (retrieval of the information). Even computer memory employs exactly these three processes. If we're going to look for the existence of even the simplest memories in plants, these are the processes we need to see happening.

The Short-Term Memory of the Venus Flytrap

As we saw back in chapter three, the Venus flytrap needs to know when an ideal meal is crawling across its leaves. Closing its trap requires a huge expense of energy, and reopening the trap can take several hours, so *Dionaea* only wants to spring closed when it's sure that the dawdling insect visiting its surface is large enough to be worth its time. The large black hairs on their lobes allow the Venus flytraps to literally feel their prey, and they act as triggers that spring the trap closed when the proper prey makes its way across the trap. If the insect touches just one hair, the trap will not spring shut; but a large enough bug will likely touch two hairs within about twenty seconds, and that signal springs the Venus flytrap into action.

We can look at this system as analogous to short-term memory. First, the flytrap encodes the information (forms the memory) that something (it doesn't know what) has touched one of its hairs. Then it stores this information for a number of seconds (retains the memory) and finally retrieves this information (recalls the memory) once a second hair is touched. If a small ant takes a while to get from one hair to the next, the trap will have forgotten the first touch by the time the ant brushes up against the next hair. In other words, it loses the storage of the information, doesn't close, and the ant happily meanders on. How does the plant encode and store the information from the unassuming bug's encounter with the first hair? How does it remember the first touch in order to react upon the second?

Scientists have been puzzled by these questions ever since John Burdon-Sanderson's early report on the physiology of the Venus flytrap in 1882. A century later, Dieter Hodick and Andreas Sievers at the University of Bonn in Germany proposed that the flytrap stored information regarding how many hairs have been touched in the electric charge of its leaf. Their model is quite elegant in its simplicity. In their studies, they discovered that touching a trigger hair on the Venus flytrap causes an electric action potential that induces calcium channels to open in the trap (this coupling of action potentials and the opening of calcium channels is similar to the processes that occur during communication between human neurons), thus causing a rapid increase in the concentration of calcium ions.

They proposed that the trap requires a relatively high concentration of calcium in order to close and that a single action potential from just one trigger hair being touched does not reach this level. Therefore, a second hair needs to be stimulated to push the calcium concentration over this threshold and spring the trap. The encoding of the information is in the initial rise in calcium levels. The retention of the information requires maintaining a high enough level of calcium so that a second increase (triggered by touching the second hair) pushes the total concentration of calcium over the threshold. As the calcium ion concentrations dissipate over time, if the second touch and potential don't happen quickly, the final concentration after the second trigger won't be high enough to close the trap, and the memory is lost.

Subsequent research supports this model. Alexander Volkov and his colleagues at Oakwood University in Alabama first demonstrated that it is indeed electricity that causes the Venus flytrap to close. To test the model they rigged up very fine electrodes and applied an electrical current to the open lobes of the trap. This made the trap close without any direct touch to its trigger hairs (while they didn't measure calcium levels, the current likely led to increases). When they modified this experiment by altering the amount of electrical current, Volkov could determine the exact electrical charge needed for the trap to close. As long as fourteen microcoulombs—a tiny bit more than the static electricity generated by rubbing two balloons together—flowed between the two electrodes, the trap closed. This could come as one large burst or as a series of smaller charges within twenty seconds. If it took longer than twenty seconds to accumulate the total charge, the trap would remain open.

Here, then, lies the proposed mechanism of the short-term memory in the Venus flytrap. The first touch of a hair activates an electric potential that radiates from cell to cell. This electric charge is stored as an increase in ion concentrations for a short time until it dissipates within about twenty seconds. But if a second action potential reaches the midrib within this time, the cumulative charge and ion concentrations pass the threshold and the trap closes. If too much time elapses between action potentials, then the plant forgets the first one, and the trap stays open.

This electric signal in the Venus flytrap (and the electric signals in other plants for that matter) are similar to the electric signals in neurons in humans and indeed all animals. The signal in both neurons and *Dionaea* leaves can be inhibited by drugs that block the ion channels which open in the membranes as the electric signal passes through the cell. When Volkov pretreated his plants with a chemical that inhibits potassium channels in human neurons, for example, the traps didn't close when they were touched or when they received the electric charges.

Long-Term Memory of Trauma

In the mid-twentieth century, some rather obscure work was carried out by the Czech botanist Rudolf Dostál, who studied what he termed "morphogenetic memory" in plants. Morphogenetic memory is a type of memory that later influences the shape or form of the plant. In other words, a plant can experience a stimulus at some point, like a rip in its leaf or a fracture of a branch, and be unaffected by it at first, but when environmental conditions change, the plant may remember the past experience and respond by changing its growth.

Dostál's experiments on flax seedlings illustrate what he meant by morphogenetic memory. To fully appreciate Dostál's experiments in this area, we have to understand a bit more about plant anatomy. Flax seedlings emerge from the ground with two large leaves called cotyledons. In the center of the two cotyledons is what's called an apical bud, which grows up from the central stem of the plant. As this bud grows up, two lateral buds

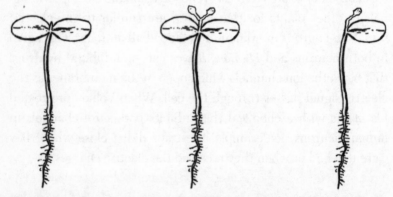

This illustration shows three flax (*Linum usitatissimum*) seedlings. The image on the left shows a two-week-old seedling with two cotyledons and an apical bud (the small bump between the two leaves). The middle picture shows a similar seedling but after the apical bud has been decapitated and the two lateral buds have been growing for about a week. The picture on the right shows a seedling with the left cotyledon removed prior to decapitating the apical bud.

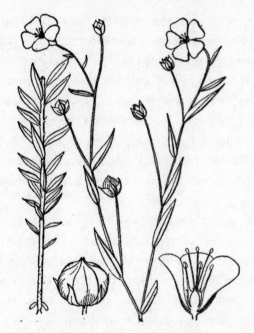

Flax (*Linum usitatissimum*)

emerge below it on each side, each facing toward one leaf. The lateral buds are dormant—they don't grow—under normal conditions. However, if the apical bud is damaged or cut off, then the lateral ones will start to grow and extend, and each one forms a new branch where each lateral bud becomes the apical bud. This repression of the lateral buds by the apical bud is called apical dominance, and releasing this repression is the basis of plant pruning. When you see a gardener pruning the hedges in front of a house, he is actually—if he's pruning correctly—removing the apical buds from each branch, encouraging more lateral buds, and new branches, to grow.

Under normal conditions, if the apical bud is pruned off, both lateral buds grow evenly. But Dostál noticed that if he removed one of the cotyledons prior to decapitation of the apical bud, the only lateral bud that would grow was the one near the remaining leaf. This result may seem like a classic case of a stimulus followed by a response. But here's where things get really interesting. When Dostál repeated the experiment and illuminated the plant with red light, the lateral bud closest to the *absent* cotyledon grew, which revealed that each bud retains the potential to grow.

Dostál's research was picked up by Michel Thellier at the University of Rouen in upper Normandy. Thellier, a member of the French Academy of Sciences, noticed that after he decapitated the apical bud on his plant of choice, *Bidens pilosa* (also known as Spanish needle), both lateral buds started to grow more or less evenly. But if he simply wounded one of the cotyledons, then only the lateral bud closest to the healthy leaf would grow. Thellier didn't have to mangle the cotyledon to get the response; he would just prick the leaf four times with a needle at the same time as the decapitation, and this minor wound was enough to get asymmetrical growth of the lateral buds.

So where does plant memory come in in what appears to be another classic stimulus-response phenomenon? Well, sometimes during these experiments Thellier would extend the amount of time between wounding the leaf and decapitating the main bud—even up to two weeks. And, lo and behold, the lateral bud farthest from the pricked cotyledon would grow out, and not both lateral buds. Thellier knew there had to be some way that the Spanish needle stored this "traumatic" experience and had a mechanism for recalling it once the central bud was removed, even if that happened many days later.

Spanish needle (*Bidens pilosa*)

The following experiment really sealed the idea that the bud of the Spanish needle remembered which of its neighboring leaves had been damaged. This time, Thellier stabbed one of the cotyledons as he had before, but then he removed *both* cotyledons several minutes later. He found that the plant retained the memory of the stabbing: once the central bud was removed, the lateral bud opposite the original wounded cotyledon grew more than the one on the side of the wounded cotyledon. The jury is still out on how this information is stored in the central bud, but one promising option is that the signal is somehow connected to auxin—the same hormone that we met in chapter five.

The Big Chill

Trofim Denisovich Lysenko was notorious for his impact on science in the Soviet Union. He rejected classic Mendelian genetics (based on the principle that all characteristics are the result of inherited genes) while championing the idea that the environment leads to the development of adaptive characteristics (such as blindness in moles living in constant darkness) that can be passed on to successive generations. This evolutionary theory, originally put forward by the noted French naturalist Jean-Baptiste Lamarck in the early nineteenth century, fit perfectly with the prevailing ideology at the time of Lysenko's research, which held that the proletariat could be modified by the environment. The Soviet establishment was so enamored of Lysenko that from 1948 to 1964 it was illegal in the Soviet Union to express any dissent from his theories. Politics aside, Lysenko made a landmark discovery in 1928 that influences plant biology to this day.

Soviet farmers grow what is called winter wheat—wheat that is planted in the fall, sprouts before freezing temperatures in the winter, and becomes dormant until the soil warms in the spring, when it flowers. Winter wheat isn't able to flower and subsequently produce grain in the spring unless it experiences a period of cold weather in the winter. The late 1920s were disastrous for Soviet agriculture because unusually warm winters destroyed most of the winter wheat seedlings—seedlings that farmers relied on to produce the grain that would feed millions.

Lysenko worked nonstop in an effort to save what little harvest they would have and to find ways of ensuring that warm winters wouldn't lead to famine in the future. He discovered that if he took winter wheat seeds and put them in a freezer before planting them, he could induce the seeds to sprout and flower without having actually endured a prolonged winter. In this way, he enabled farmers to plant wheat in the springtime, and he ultimately saved wheat yields in his country. Lysenko called the process "vernalization," which has been embraced now as the general term for any cold treatment, be it natural or artificial.

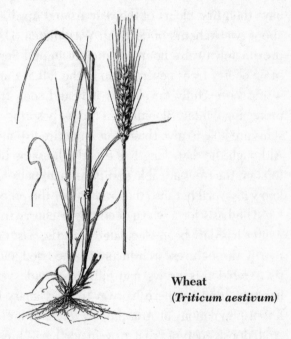

Wheat
(*Triticum aestivum*)

Other scientists also knew that some plants needed cold weather in order to flower (one of the first reports came out of the Ohio Board of Agriculture in 1857), but Lysenko was the first to show that the process could be artificially manipulated. Many plants rely on the cold temperatures of winter to cultivate their harvest; many fruit trees will only flower and set fruit following a cold winter, and lettuce and arabidopsis seeds only germinate following a cold snap. The ecological advantage of vernalization is clear: it ensures that following the cold of winter, a plant will sprout or flower in the spring or summer, and not during other times of the year when the amount of light and temperature could also support plant growth.

For example, the cherry trees in Washington, D.C., usually have their first bloom of the year around April 1, when there are about twelve hours of daylight. Washington, D.C., also sees approximately twelve hours of sunlight in mid-September, but the same cherry trees never bloom in the fall; if they did, their fruit would never fully develop as it would soon freeze in the approaching winter. Blooming in the early spring, the cherry blossoms are able to give their fruit an entire five months to mature. Although the day's length is exactly the same in April and September, the trees are able to differentiate between the two. They know it's April because they remember the preceding winter.

The basis for a wheat seedling or cherry tree remembering winter has only been elucidated over the past decade or so, primarily through research involving the tried-and-true arabidopsis. Arabidopsis grows naturally in a wide variety of natural habitats, from northern Norway to the Canary Islands. The different populations of *Arabidopsis thaliana* are called ecotypes. Arabidopsis ecotypes that grow in northern climates need vernalization to flower, while those that grow in warmer climates do just fine without it. This need for vernalization is encoded in the genes of the northern ecotypes. If you cross a plant that needs winter in order to flower with a plant that doesn't, the offspring still need a cold snap in order to flower; genetically, the need for cold is a dominant trait (just as brown eyes is a dominant trait relative to blue eyes in people). The specific gene involved is

FLC, which stands for *flowering locus C*. In its dominant version, *FLC* inhibits flowering until the plant has undergone a vernalization period.

Once the plant goes through a period of cold weather, the *FLC* gene is no longer transcribed; the gene is turned off. But that doesn't mean the plants will immediately start to flower; it only means that the plants *could* flower if other conditions, such as light and temperature, cooperate. So the plant must have a way of remembering that it once experienced a cold climate to keep *FLC* turned off, even though temperatures have since warmed up.

Many researchers have tried to understand just how vernalization turns off *FLC* and how it stays this way once it's turned off. These investigations have highlighted how epigenetics is intertwined with a plant's memory of winter. Epigenetics refers to changes in gene activity that don't require alterations in the DNA *code*, as mutations do, yet these changes in gene activity are still passed down from parent to offspring.* In many cases, epigenetics works through changes in the structure of the DNA.

In cells, DNA is organized in chromosomes, which are much more than simple strings of nucleotides. The double helix of DNA wraps around proteins called histones, forming what is known as chromatin. This chromatin can twist even more, just like an overly twisted rubber band, compacting the DNA and proteins into highly condensed and packed structures. These structures are dynamic: different parts of the chromatin can unravel or pack up again. Active genes (those that are transcribed) are found in

---

*Epigenetics encompasses a wide range of heritable changes that are independent of DNA sequence. These include chemical changes in histones, chemical changes of the DNA (for example, DNA methylation—see p. 128), different types of small RNAs, and infectious proteins known as prions.

areas of the chromatin that are unraveled, while inactive genes lie in regions that are more condensed.*

The histone proteins are one of the key factors that determine how tightly knit the chromatin gets, and this is very important for understanding how *FLC* is activated. Scientists have discovered that cold treatment triggers a change in the structure of the histones (a process called methylation) around the *FLC* gene, which enables the chromatin to be tightly packed. This turns off *FLC*, and the plant is able to flower. This epigenetic change (the type of histone around the gene) is passed down from parent to daughter cells over successive generations, and the *FLC* gene remains inactive in all cells even *after* the cold weather subsides. Once the *FLC* gene has been turned off, the plants can wait until the rest of the environmental conditions are ideal for flowering. In perennial plants like oak trees and azaleas, which flower once per year, the *FLC* gene has to be reactivated once the plant has blossomed to inhibit promiscuous flowering that might occur out of season until the next winter has passed. This involves the cells reprogramming their histone code, which opens the chromatin around the *FLC* gene, reactivating it. How this occurs, and how it's regulated, are matters of current research.

This epigenetic mechanism of cellular memory is not specific to plants and is the basis of a great many biological processes and diseases. Epigenetics has caused a paradigm shift in biology because it goes against the classic genetic concept that the only changes that can be passed on from cell to cell are those in the DNA sequence. What's truly amazing about epigenetics is that it

*A major differentiator of cell types, like blood cells versus liver cells in people, or pollen versus leaf cells in plants, is the structure of their chromatin, which affects which genes are activated.

facilitates memory not only from season to season within a single organism but from generation to generation.

In Every Generation . . .

Memories are actively handed down from one generation to the next through rituals, storytelling, and more. But transgenerational memory involving epigenetics is completely different. This type of memory usually involves information about an environmental or physical stress that's passed down from parents to offspring. Barbara Hohn's laboratory in Basel, Switzerland, was the first to provide evidence for such transgenerational memory. Hohn and her colleagues knew that conditions that create stress on a plant, like ultraviolet light or pathogen attack, lead to changes in the plant genome that result in new combinations of DNA.

These stress-induced changes make sense ecologically because—like any other organism—a plant needs to find ways to survive under stress. One of the ways a plant does this is through new genetic variations. Hohn's astounding study showed that not only do the stressed plants make new combinations of DNA but their offspring also make the new combinations, even though they themselves had never been directly exposed to any stress. The stress in the parents caused a stable heritable change that was passed on to all their offspring: the plants behaved as if they'd been stressed. They remembered that their parents had been through this stress and reacted similarly.

This use of the word "remembered" may seem unorthodox, but let's analyze this in light of the three steps of memory we encountered at the beginning of this chapter: the parents formed the memory of the stress, retained it, and passed it on to their children, and the children recalled the information and reacted accordingly, in this case, with increased genomic changes.

The implications of this study are vast. An environmental stress causes a heritable change that is passed on to successive generations. This fits excellently with the theories of Jean-Baptiste Lamarck, who, as you may recall, claimed that evolution was based on the inheritance of acquired characteristics. Hohn's plants, following the UV or pathogen stress, acquired the characteristic of increased genetic variation and passed it on to all of their progeny (and a single arabidopsis plant produces thousands of seeds!). This cannot be explained by mutations in the DNA sequence of the stressed plants, because this could at most be passed on to only a very small percentage of the progeny. On the other hand, if the stress induced an epigenetic change, this could happen in *all* the cells at once, including pollen and egg cells, and be passed on to the entire next generation, as well as many future ones. While scientists speculate as to the nature of the epigenetic change involved in these memories, it remains undiscovered.

Igor Kovalchuk created a follow-up study in which he included other stresses on genetic variation in plants and their progeny including heat and salt. He showed that these different environmental insults increase the frequency of genomic rearrangements not only in the parental generation but also in the second generation. Kovalchuk's results were fascinating because they revealed even more than this. Not only did the second generation of plants show increased genetic variation, confirming Hohn's results, but they were also more tolerant to the various stresses. In other words, stressed parents gave rise to offspring that grew better under harsh conditions compared with regular plants. The various stresses almost certainly induce epigenetic changes in chromatin structure in the parents, which they pass on to their progeny. We believe this because Kovalchuk's group showed that if they treated the offspring with a chemical that wiped out epigenetic information, these same plants lost their ability to thrive under the environmental stress. Hohn's results were not universally accepted, as is the case with many paradigm-shifting studies in science. The growing consensus, however, is that her results, as well as others, have heralded a new era in ge-netics. The idea of stress leading to memories that are passed down from one generation to the next is supported by an increas-ing number of studies, not only in plants, but in animals as well. In all cases, this "memory" is based on some form of epigenetic heredity.

Intelligent Memory?

Plants clearly have the ability to store and recall biological infor-mation. Intuitively, we know that this is quite different from the detailed and emotion-filled memories we recall every day. But at a basic level, the behaviors of different plants described in this chapter are remedial types of memory. The tendril's coiling, the Venus flytrap's closing, and the arabidopsis's remembering envi-ronmental stress all include the processes of forming the mem-ory of the event, retaining the memory for distinct time periods, and recalling the memory at a later point in order to get a specific developmental response.

Many of the mechanisms involved in plant memory are also involved in human memory, including epigenetics and electro-chemical gradients. These gradients are the bread and butter of neural connections in our brains, the seat of memory as most of us understand it. Over the past several years, plant scientists have discovered that not only do plant cells communicate with electrical currents (as we saw in several chapters) but plants also contain proteins known in humans and other animals as neuroreceptors. A perfect example is the glutamate receptor. Glutamate receptors in the brain are very important for neural communication, memory formation, and learning, and a number of neuroactive drugs target glutamate receptors. It was a great surprise, then, for scientists at New York University to discover that plants contain glutamate receptors and that arabidopsis plants are sensitive to neuroactive drugs that alter glutamate receptor activity. At this point we still don't fully understand what glutamate receptors do in plants, but very recent work carried out by José Feijó and his group in Portugal shows that these receptors in plants function in cell-to-cell signaling in a way that's very similar to how human neurons communicate with each other. This leaves us to marvel at the evolutionary role of "brain receptors" in plants. Perhaps the similarities between human brain function and plant physiology may be greater than we've assumed.

Plant memories, like human immune memory, are not semantic or episodic memories, as Tulving defined them, but rather procedural memories, memories of how to do things; these memories depend on the ability to sense external stimulation. Tulving further proposed that each of the three levels of memory is associated with an increasing level of consciousness. Procedural memory is associated with anoetic consciousness, semantic memory is associated with noetic consciousness, and episodic memory is associated with autonoetic consciousness. Plants clearly do not fit the definition of consciousness associated with semantic or episodic memories. But as stated in a recent opinion article, "The lowest level of consciousness characteristic for procedural memory—anoetic consciousness—refers to the ability of organisms to sense and to react to external and internal stimulation, which all plants and simple animals are capable of." This leads us to the most intriguing question of all: If plants exhibit different types of memory and have a form of consciousness, should they be considered intelligent?

Epilogue: The Aware Plant

"Intelligence" is a loaded term. Everyone from Alfred Binet, the inventor of the much-debated IQ test, to the renowned psychologist Howard Gardner has had a different understanding of just what it means to classify someone as "intelligent." While some researchers consider intelligence a unique propensity of human beings, we have seen reports that animals—from orangutans to octopuses—possess qualities that fall under some definitions of "intelligence." Applying definitions of intelligence to plants, however, is much more contentious, though the question of intelligent plants is hardly a new one. Dr. William Lauder Lindsay, who doubled as a physician *and* a botanist, wrote in 1876: "It appears to me that *certain attributes of mind, as it occurs in Man, are common to Plants.*"

Anthony Trewavas, an esteemed plant physiologist based at the University of Edinburgh in Scotland and one of the early modern purveyors of plant intelligence, points out that while humans are clearly more intelligent than other animals, it is unlikely that intelligence as a biological property originated only in *Homo sapiens*. In this vein he sees intelligence as a biological characteristic no different from, say, body shape and respiration—all of which evolved through the natural selection of characteristics present in earlier organisms. We saw this quite clearly in Chapter Four in the "deaf" genes shared by plants and humans. These genes were present in a common ancient ancestor of plants and animals, and Trewavas has proposed that rudimentary intelligence was present there as well.

Controversy arose among plant biologists when a group of scientists who study various aspects of plant function defined a new field in 2005 that they called "plant neurobiology," which aims to study the information networks present in plants. These scientists saw many similarities between plant anatomy and physiology and the neural networks in animals. Some of these similarities are obvious, such as the electrical signaling we encountered in the Venus flytrap and *Mimosa* plants, and some more divisive, such as the architecture of plant roots being similar to the architecture of neural networks found in various animals.

This latter hypothesis was originally put forward in the nineteenth century by Charles Darwin and has been picked up again over the past few years, particularly by Stefano Mancuso from the University of Florence and Frantiöek Baluöka from the University of Bonn, two of the pioneers in the field of plant neurobiology. Many other biologists who study plants, including a number of very prominent scientists, criticized the ideas behind plant neurobiology, claiming that its theoretical basis is flawed and that it has not added to our understanding of plant physiology or plant cell biology. They strongly felt that the plant neurobiologists had gone too far in drawing parallels between plant and animal biology.

Many proponents of plant neurobiology would be the first to explain that the term itself is provocative and therefore useful for encouraging more debate and discussion about the parallels between the ways plants and animals process information. Metaphors, as pointed out by Trewavas and others, help us make connections that we might not normally make. If by using the term "plant neurobiology" we challenge people to reevaluate their understanding of biology in general, and plant biology specifically, then the term is valid. But we must be clear: no matter what similarities we may find at the genetic level between plants and animals (and, as we have seen, they are significant), they are two very unique evolutionary adaptations for multicellular life, each of which depends on unique kingdom-specific sets of cells, tissues, and organs. For example, vertebrate animals developed a bony skeleton to support weight, while plants developed a woody trunk. Both fill similar functions, yet each is biologically unique.[*]

While we could subjectively define "vegetal intelligence" as another facet of multiple intelligences, such a definition does not further our understanding of either intelligence or plant biology. The question, I posit, should not be whether or not plants are *intelligent*—it will be ages before we all agree on what that term means; the question should be, "Are plants aware?" and, in fact, they are. Plants are acutely aware of the world around them. They are aware of their visual environment; they differentiate between red, blue, far-red, and UV lights and respond accordingly. They are aware of aromas surrounding them and respond to minute quantities of volatile compounds wafting in the air. Plants know when they are being touched and can distinguish different touches. They are aware of gravity: they can change their shapes to ensure that shoots grow up and roots grow down. And plants are aware of their past: they remember past infections and the

---

[*]Following the initial controversy, and after much discussion, the plant neurobiologists changed the name of their professional organization in 2009 from the Society for Plant Neurobiology to the more accepted Society of Plant Signaling and Behavior (though "behavior" is also an interesting term not intuitively connected to plants).

conditions they've weathered and then modify their current physiology based on these memories.

If a plant is aware, what does this mean for us regarding our own interactions with the green world? On one hand, an "aware plant" is not aware of us as individuals. We are simply one of many external pressures that increase or decrease a plant's chances for survival and reproductive success. To borrow terms from Freudian psychology: the plant psyche is devoid of an ego and a superego, though it may contain an id, the unconscious part of the psyche that gets sensory input and works according to instinct. A plant is aware of its environment, and people are part of this environment. But it's not aware of the myriad gardeners and plant biologists who develop what they consider to be personal relationships with their plants. While these relationships may be meaningful to the caretaker, they are not dissimilar to the relationship between a child and her imaginary friend; the flow of meaning is unidirectional. I've heard world-famous scientists and undergraduate research students alike use anthropomorphic language with abandon as they describe their plants as "not looking too happy" when mildew has taken over their leaves or as "satisfied" after they've been watered.

These terms represent our own subjective assessment of a plant's decidedly unemotional physiological status. For all the rich sensory input that plants and people perceive, only humans render this input as an emotional landscape. We project on plants our emotional load and assume that a flower in full bloom is happier than a wilting one. If "happy" can be defined as an "optimal physiological state," then perhaps the term fits. But I think that for all of us, "happy" depends on much more than being in perfect physical health. In fact, we've all known people afflicted with various ailments who considered themselves happy, and otherwise healthy individuals who are generally miserable in mood. Happiness, we can agree, is a state of mind.

A plant's awareness also does not imply that a plant can suffer. A seeing, smelling, feeling plant can no more suffer pain than can a computer with a faulty hard drive. Indeed, "pain" and "suffering," like "happy," are very subjective terms and are out of place when describing plants. The International Association for the Study of Pain defines pain as "an unpleasant sensory and emotional experience associated with actual or potential tissue damage, or described in terms of such damage." Perhaps "pain" for a plant could be defined in terms of "actual or potential tissue damage," as when a plant senses physical distress that can lead to cell damage or death. A plant senses when a leaf has been punctured by an insect's jaws; a plant knows when it's been burned in a forest fire. Plants know when they lack water during a drought. But plants do not suffer. They don't have, to our current knowledge, the capacity for an "unpleasant . . . emotional experience." Indeed, even in humans, pain and suffering are considered separate phenomena, interpreted by different parts of the brain. Brain-imaging studies have identified pain centers deep within the human brain that radiate out from the brain stem, while the capacity for suffering, scientists believe, is located in the prefrontal cortex. So if suffering from pain necessitates highly complex neural structures and connections of the frontal cortex, which are present only in higher vertebrates, then plants obviously don't suffer: they have no brain.

The construct of a brainless plant is important for me to accentuate. If we keep in mind that a plant doesn't have a brain, it follows then that any anthropomorphic description is at its base severely limited. It allows us to continue to anthropomorphize plant behavior for the sake of literary clarity while remembering that all such descriptions must be tempered by the idea of a brainless plant. While we use the same terms—"see," "smell," "feel"—we also know that the overall sensual experience is qualitatively different for plants and people.

Without this caveat, anthropomorphism of plant behavior left unchecked can lead to unfortunate, if not humorous, consequences. For example, in 2008 the Swiss government established an ethics committee to protect the "dignity" of plants.* A brainless plant likely does not worry about its dignity. And yet if a plant is aware, it means everything for us regarding our own interactions with the vegetal world. Maybe the Swiss attempt at bestowing dignity on plants mirrors our own attempt at defining our relationship with the plant world. As individuals, we often seek our place in society by comparing ourselves with other people. As a species, we seek our place in nature by comparing ourselves with other animals. It's easy for us to see ourselves in the eyes of a chimpanzee; we can identify with a baby gorilla clinging to her mother. John Grogan's dog, Marley, like Lassie and Rin Tin Tin before him, evokes very deep feelings of empathy, and even people who are not necessarily dog lovers can see human characteristics in our canine friends. I've known bird keepers who claim their parrots understand them, and fish lovers who see human behavior in their marine life. These examples clearly show that "human" may be only a flavor, albeit an interesting one, of intelligence.

So if humans and plants are similar in that both are aware of complex light environments, intricate aromas, different physical stimulations, if humans and plants both have preferences, and if

---

*This committee was formed to further define dignity in terms of plants, as the Swiss Federal Constitution requires "account to be taken of the dignity of living beings when handling animals, plants and other organisms." See www.ekah.admin.ch/en/topics/dignity -of-living-beings/index.html.

both remember, then do we see ourselves when looking at the plant?

What we must see is that on a broad level we share *biology* not only with chimps and dogs but also with begonias and sequoias. We should see a very long-lost cousin when we gaze at our rosebush in full bloom, knowing that we can discern complex environments just as it can, knowing that we share common genes. When we look at ivy clinging to a wall, we are looking at what, save for some ancient stochastic event, could have been our fate. We are seeing another possible outcome of our own evolution, one that branched off some two billion years ago.

A shared genetic past does not negate eons of separate evolution. While plants and humans maintain parallel abilities to sense and be aware of the physical world, the independent paths of evolution have led to a uniquely human capacity, beyond intelligence, that plants don't have: the ability to care.

So the next time you find yourself on a stroll through a park, take a second to ask yourself: What does the dandelion in the lawn see? What does the grass smell? Touch the leaves of an oak, knowing that the tree will remember it was touched. But it won't remember you. You, on the other hand, can remember this particular tree and carry the memory of it with you forever.

NOTES

ACKNOWLEDGMENTS

INDEX

Notes

PROLOGUE

3 *In my research*: Daniel A. Chamovitz et al., "The COP9 Complex, a Novel Multisubunit Nuclear Regulator Involved in Light Control of a Plant Developmental Switch," *Cell* 86, no. 1 (1996): 115–21.

3 *Much to my surprise*: Daniel A. Chamovitz and Xing-Wang Deng, "The Novel Components of the Arabidopsis Light Signaling Pathway May Define a Group of General Developmental Regulators Shared by Both Animal and Plant Kingdoms," *Cell* 82, no. 3 (1995): 353–54.

3 *Many years later*: Alyson Knowles et al., "The COP9 Signalosome Is Required for Light-Dependent Timeless Degradation and *Drosophila* Clock Resetting," *Journal of Neuroscience* 29, no. 4 (2009): 1152–62; Ning Wei, Giovanna Serino, and Xing-Wang Deng, "The COP9 Signalosome: More Than a Protease," *Trends in Biochemical Sciences* 33, no. 12 (2008): 592–600.

5 *As the renowned plant physiologist*: Peter Tompkins and Christopher Bird, *The Secret Life of Plants* (New York: Harper & Row, 1973); Arthur W. Galston, "The Unscientific Method," *Natural History* 83 (1974): 18, 21, 24.

ONE. WHAT A PLANT SEES

10 *"the physical sense by which light"*: "Sight," *Merriam-Webster*, www.merriam-webster.com/dictionary/sight.

12 *"There are extremely few"*: Charles Darwin and Francis Darwin, *The Power of Movement in Plants* (New York: D. Appleton, 1881), p. 1.

13 *"could not see the seedlings"*: Ibid., p. 450.

16 *In 1918*: A brief history of light research at USDA can be found at www.ars.usda.gov/is/timeline/light.htm.

17 *This phenomenon, called photoperiodism*: Wightman W. Garner and Harry A. Allard, "Photoperiodism, the Response of the Plant to Relative Length of Day and Night," *Science* 55, no. 1431 (1922): 582–83.

18 *the plants, and it*: Marion W. Parker et al., "Action Spectrum for the Photoperiodic Control of Floral Initiation in Biloxi Soybean," *Science* 102, no. 2641 (1945): 152–55.

18 *Then, in the early 1950s*: Harry Alfred Borthwick, Sterling B. Hendricks, and Marion W. Parker, "The Reaction Controlling Floral Initiation," *Proceedings of the National Academy of Sciences of the United States of America* 38, no. 11 (1952): 929–34; Harry Alfred Borthwick et al., "A Reversible Photoreaction Controlling Seed Germination," *Proceedings of the National Academy of Sciences of the United States of America* 38, no. 8 (1952): 662–66.

19 *Warren L. Butler and colleagues*: Warren L. Butler et al., "Detection, Assay, and Preliminary Purification of the Pigment Controlling Photoresponsive Development of Plants," *Proceedings of the National Academy of Sciences of the United States of America* 45, no. 12 (1959): 1703–8.

20 *The approach spearheaded*: Maarten Koornneef, E. Rolff, and Carel Johannes Pieter Spruit, "Genetic Control of Light-Inhibited Hypocotyl Elongation in *Arabidopsis thaliana* (L) Heynh," *Zeitschrift für Pflanzenphysiologie* 100, no. 2 (1980): 147–60.

22 *To make a long*: Joanne Chory, "Light Signal Transduction: An Infinite Spectrum of Possibilities," *Plant Journal* 61, no. 6 (2010): 982–91.

24 *These eyespots have been*: Georg Kreimer, "The Green Algal Eyespot Apparatus: A Primordial Visual System and More?," *Current Genetics* 55, no. 1 (2009): 19–43.

25 *The name "cryptochrome" is actually*: Jonathan Gressel, "Blue-Light Photoreception," *Photochemistry and Photobiology* 30, no. 6 (1979): 749–54.

25 *cryptochrome is no longer*: Margaret Ahmad and Anthony R. Cashmore, "*HY4* Gene of *A. thaliana* Encodes a Protein with Characteristics of a Blue-Light Photoreceptor," *Nature* 366, no. 6451 (1993): 162–66.

26 *The plant cryptochrome*: Anthony R. Cashmore, "Cryptochromes: Enabling Plants and Animals to Determine Circadian Time," *Cell* 114, no. 5 (2003): 537–43.

TWO. WHAT A PLANT SMELLS

27 *"to perceive odor or scent"*: Available at Merriam-Webster.com.

30 *Frank E. Denny, a scientist*: Frank E. Denny, "Hastening the Coloration of Lemons," *Agricultural Research* 27 (1924): 757–69.

30 *Gane analyzed the air*: Richard Gane, "Production of Ethylene by Some Ripening Fruits," *Nature* 134 (1934): 1008; and William Crocker, A. E. Hitchcock, and P. W. Zimmerman, "Similarities in the Effects of Ethylene and the Plant Auxins," *Contributions from Boyce Thompson Institute* 7 (1935): 231–48.

33 *One of her projects centered on*: Justin B. Runyon, Mark C. Mescher, and Consuelo M. De Moraes, "Volatile Chemical Cues Guide Host Location and Host Selection by Parasitic Plants," *Science* 313, no. 5795 (2006): 1964–67.

35 *as Rhoades discovered*: David F. Rhoades, "Responses of Alder and Willow to Attack by Tent Caterpillars and Webworms: Evidence for Pheromonal Sensitivity of Willows," in *Plant Resistance to Insects*, edited by Paul A. Hedin (Washington, D.C.: American Chemical Society, 1983), pp. 55–66.

36 *Dartmouth researchers Ian Baldwin and Jack Schultz*: Ian T. Baldwin and Jack C. Schultz, "Rapid Changes in Tree Leaf Chemistry Induced by Damage: Evidence for Communication Between Plants," *Science* 221, no. 4607 (1983): 277–79.

38 *These early reports of plant signaling*: Simon V. Fowler and John H. Lawton, "Rapidly Induced Defenses and Talking Trees: The Devil's Advocate Position," *American Naturalist* 126, no. 2 (1985): 181–95.

38 *the popular press embraced the idea*: "Scientists Turn New Leaf, Find Trees Can Talk," *Los Angeles Times*, June 6, 1983, A9; "Shhh. Little Plants Have Big Ears," *Miami Herald*, June 11, 1983, 1B; "Trees Talk, Respond to Each Other, Scientists Believe," *Sarasota Herald-Tribune*, June 6, 1983; and "When Trees Talk," *New York Times*, June 7, 1983.

39 *Martin Heil and his team*: Martin Heil and Juan Carlos Silva Bueno, "Within-Plant Signaling by Volatiles Leads to Induction and Priming of an Indirect Plant Defense in Nature," *Proceedings of the National Academy of Sciences of the United States of America* 104, no. 13 (2007): 5467–72.

43 *Heil collaborated with colleagues from South Korea*: Hwe-Su Yi et al., "Airborne Induction and Priming of Plant Defenses Against a Bacterial Pathogen," *Plant Physiology* 151, no. 4 (2009): 2152–61.

45 *supported work done a decade earlier*: Vladimir Shulaev, Paul Silverman, and Ilya Raskin, "Airborne Signalling by Methyl Salicylate in Plant Pathogen Resistance," *Nature* 385, no. 6618 (1997): 718–21

45 *Plants can convert soluble salicylic acid*: Mirjana Seskar, Vladimir Shulaev, and Ilya Raskin, "Endogenous Methyl Salicylate in Pathogen-Inoculated Tobacco Plants," *Plant Physiology* 116, no. 1 (1998): 387–92.

47 *the author Michael Pollan infers*: Michael Pollan, *The Botany of Desire: A Plant's-Eye View of the World* (New York: Random House, 2001).

47 *A recent (and provocative) study*: Shani Gelstein et al., "Human Tears Contain a Chemosignal," *Science* 331, no. 6014 (2011): 226–30.

THREE. WHAT A PLANT FEELS

55 *"one of the most wonderful [plants]"*: Charles Darwin, *Insectivorous Plants* (London: John Murray, 1875), p. 286.

55 *"During the summer of 1860"*: Ibid., p. 1.

56 *"Drops of water"*: Ibid., p. 291.

56 *John Burdon-Sanderson made the crucial*: John Burdon-Sanderson, "On the Electromotive Properties of the Leaf of *Dionaea* in the Excited and Unexcited States," *Philosophical Transactions of the Royal Society* 173 (1882): 1–55.

57 *electric stimulation itself*: Alexander G. Volkov, Tejumade Adesina, and Emil Jovanov, "Closing of Venus Flytrap by Electrical Stimulation of Motor Cells," *Plant Signaling & Behavior* 2, no. 3 (2007): 139–45.

57 *Volkov's work and earlier research*: Ibid.; Dieter Hodick and Andreas Sievers, "The Action Potential of *Dionaea muscipula* Ellis," *Planta* 174, no. 1 (1988): 8–18.

58 *This was noticed by*: Virginia A. Shepherd, "From Semi-conductors to the Rhythms of Sensitive Plants: The Research of J. C. Bose," *Cellular and Molecular Biology* 51, no. 7 (2005): 607–19.

59 *Unfortunately, Burdon-Sanderson was*: Subrata Dasgupta, "Jagadis Bose, Augustus Waller, and the Discovery of 'Vegetable Electricity,'" *Notes and Records of the Royal Society of London* 52, no. 2 (1998): 307–22.

61 *"We were confronted with"*: Frank B. Salisbury, *The Flowering Process*, International Series of Monographs on Pure and Applied Biology, Division: Plant Physiology (New York: Pergamon Press, 1963).

62 *He coined the cumbersome*: Mark J. Jaffe, "Thigmomorphogenesis: The Response of Plant Growth and Development to Mechanical Stimulation— with Special Reference to *Bryonia dioica*," *Planta* 114, no. 2 (1973): 143–57.

64 *Braam understood that*: Janet Braam and Ronald W. Davis, "Rain-Induced, Wind-Induced, and Touch-Induced Expression of Calmodulin and Calmodulin-Related Genes in Arabidopsis," *Cell* 60, no. 3 (1990): 357–64.

66 *Thanks to the continuing work*: Dennis Lee, Diana H. Polisensky, and Janet Braam, "Genome-Wide Identification of Touch- and Darkness-Regulated Arabidopsis Genes: A Focus on Calmodulin-Like and *XTH* Genes," *New Phytologist* 165, no. 2 (2005): 429–44.

67 *An amazing example of*: David C. Wildon et al., "Electrical Signaling and Systemic Proteinase-Inhibitor Induction in the Wounded Plant," *Nature* 360, no. 6399 (1992): 62–65.

FOUR. WHAT A PLANT HEARS

72 *Many of us have heard*: For example, "Plants and Music," www.miniscience.com/projects/plantmusic/index.html.

72 *Typically, though, much*: Ross E. Koning, Science Projects on Music and

Sound, Plant Physiology Information website, http://plantphys.info/music
.shtml; www.youth.net/nsrc/sci/sci048.html#anchor992130; http://jrscience
.wcp.muohio.edu/nsfall05/LabpacketArticles/Whichtypeofmusicbest
stimu.html; http://spider2.allegheny.edu/student/S/sesekj/FS%20Bio%202
01%20Coenen%20Draft%20Results-Discussion.doc.

72 *A common definition of "hearing"*: Hearing Impairment Information,
www.disabled-world.com/disability/types/hearing.

74 *"fool's experiment"*: Francis Darwin, ed., *Charles Darwin: His Life Told
in an Autobiographical Chapter and in a Selected Series of His Published Letters* (London: John Murray, 1892).

74 *An example of one*: Katherine Creath and Gary E. Schwartz, "Measuring
Effects of Music, Noise, and Healing Energy Using a Seed Germination
Bioassay," *Journal of Alternative and Complementary Medicine* 10, no. 1
(2004): 113–22.

74 *Schwartz founded the VERITAS*: The Veritas Research Program, http://
veritas.arizona.edu.

74 *Obviously, studying consciousness after*: Ray Hyman, "How Not to Test
Mediums: Critiquing the Afterlife Experiments," www.csicop.org/si/show
/how_not_to_test_mediums_critiquing_the_afterlife_experiments//; Robert Todd Carroll, "Gary Schwartz's Subjective Evaluation of Mediums:
Veritas or Wishful Thinking?," http://skepdic.com/essays/gsandsv.html.

74 *"biologic effects of music"*: Creath and Schwartz, "Measuring Effects of
Music, Noise, and Healing Energy."

75 *who used ultrasonic waves*: Pearl Weinberger and Mary Measures, "The
Effect of Two Audible Sound Frequencies on the Germination and Growth
of a Spring and Winter Wheat," *Canadian Journal of Botany* 46, no. 9
(1968): 1151–58. Pearl Weinberger and Mary Measures, "Effects of the
Intensity of Audible Sound on the Growth and Development of Rideau
Winter Wheat," *Canadian Journal of Botany* 57, no. 9 (1979): 1151036–39.

75 *"doctor's wife, housekeeper"*: Dorothy L. Retallack, *The Sound of Music
and Plants* (Santa Monica, Calif.: DeVorss, 1973).

75 *she enrolled as a freshman*: Anthony Ripley, "Rock or Bach an Issue to
Plants, Singer Says," *New York Times*, February 21, 1977.

76 *Retallack explained that she*: Franklin Loehr, *The Power of Prayer on
Plants* (Garden City, N.Y.: Doubleday, 1959).

77 *Retallack's studies were fraught*: Linda Chalker-Scott, "The Myth of
Absolute Science: 'If It's Published, It Must Be True,'" www.puyallup
.wsu.edu/~linda%20chalker-scott/horticultural%20myths_files/Myths
/Bad%20science.pdf.

78 *contradicted an important study*: Richard M. Klein and Pamela C. Edsall, "On the Reported Effects of Sound on the Growth of Plants," *Biosci-
ence* 15, no. 2 (1965): 125–26.

79 *"There was no leaf abscission"*: Ibid.

80 *It is also featured*: Peter Tompkins and Christopher Bird, *The Secret Life of Plants* (New York: Harper & Row, 1973).

80 *"The trouble with* The Secret Life*"*: Arthur W. Galston, "The Unscientific Method," *Natural History* 83, no. 3 (1974): 18, 21, 24.

80 *In Janet Braam's original*: Janet Braam and Ronald W. Davis, "Rain-Induced, Wind-Induced, and Touch-Induced Expression of Calmodulin and Calmodulin-Related Genes in Arabidopsis," *Cell* 60, no. 3 (1990): 357–64.

81 *Similarly, in* Physiology: Peter Scott, *Physiology and Behaviour of Plants* (Hoboken, N.J.: John Wiley, 2008).

83 *More than three hundred researchers*: The Arabidopsis Genome Initiative, "Analysis of the Genome Sequence of the Flowering Plant *Arabidopsis thaliana*," *Nature* 408, no. 6814 (2000): 796–815.

84 *Most relevant to our*: Alan M. Jones et al., "The Impact of *Arabidopsis* on Human Health: Diversifying Our Portfolio," *Cell* 133, no. 6 (2008): 939–43.

84 *The human genome*: Daniel A. Chamovitz and Xing-Wang Deng, "The Novel Components of the Arabidopsis Light Signaling Pathway May Define a Group of General Developmental Regulators Shared by Both Animal and Plant Kingdoms," *Cell* 82, no. 3 (1995): 353–54.

85 *mutations in the arabidopsis breast cancer*: Kiyomi Abe et al., "Inefficient Double-Strand DNA Break Repair Is Associated with Increased Fascination in *Arabidopsis* BRCA2 Mutants," *Journal of Experimental Botany* 70, no. 9 (2009): 2751–61.

86 *When a mutation occurs*: Valera V. Peremyslov et al., "Two Class XI Myosins Function in Organelle Trafficking and Root Hair Development in Arabidopsis," *Plant Physiology* 146, no. 3 (2008): 1109–16.

86 *Professor Stefano Mancuso*: "Phonobiologic Wines," www.brightgreencities .com/v1/en/bright-green-book/italia/vinho-fonobiologico.

87 *In a similar vein, Roman*: Roman Zweifel and Fabienne Zeugin, "Ultrasonic Acoustic Emissions in Drought-Stressed Trees—More Than Signals from Cavitation?," *New Phytologist* 179, no. 4 (2008): 1070–79.

88 *The great evolutionary biologist*: Theodosius Dobzhansky, "Biology, Molecular and Organismic," *American Zoologist* 4, no. 4 (1964): 443–52.

FIVE. HOW A PLANT KNOWS WHERE IT IS

94 *Henri-Louis Duhamel du Monceau*: Henri-Louis Duhamel du Monceau, *La physique des arbres où il est traité de l'anatomie des plantes et de l'économie végétale: Pour servir d'introduction au "Traité complet des bois & des forests," avec une dissertation sur l'utilité des méthodes de*

botanique & une explication des termes propres à cette science & qui sont en usage pour l'exploitation des bois & des forêts (Paris: H. L. Guérin & L. F. Delatour, 1758).

95 *"the hypothesis"*: Thomas Andrew Knight, "On the Direction of the Radicle and Germen During the Vegetation of Seeds," *Philosophical Transactions of the Royal Society of London* 96 (1806): 99–108.

97 *As with so many questions*: Charles Darwin and Francis Darwin, *The Power of Movement in Plants* (New York: D. Appleton, 1881).

99 *More than a century*: Ryuji Tsugeki and Nina V. Fedoroff, "Genetic Ablation of Root Cap Cells in *Arabidopsis*," *Proceedings of the National Academy of Sciences of the United States of America* 96, no. 22 (1999): 12941–46.

99 *many scientists have isolated*: Miyo Terao Morita, "Directional Gravity Sensing in Gravitropism," *Annual Review of Plant Biology* 61 (2010): 705–20.

100 *For example, an arabidopsis*: Joanna W. Wysocka-Diller et al., "Molecular Analysis of SCARECROW Function Reveals a Radial Patterning Mechanism Common to Root and Shoot," *Development* 127, no. 3 (2000): 595–603.

100 *Recent genetic studies demonstrate*: Daisuke Kitazawa et al., "Shoot Circumnutation and Winding Movements Require Gravisensing Cells," *Proceedings of the National Academy of Sciences of the United States of America* 102, no. 51 (2005): 18742–47.

101 scarecrow *gene is necessary*: Wysocka-Diller et al., "Molecular Analysis of SCARECROW Function."

103 *Using a high-gradient*: Sean E. Weise et al., "Curvature in *Arabidopsis* Inflorescence Stems Is Limited to the Region of Amyloplast Displacement," *Plant and Cell Physiology* 41, no. 6 (2000): 702–9.

103 *Under these weightless conditions*: John Z. Kiss, W. Jira Katembe, and Richard E. Edelmann, "Gravitropism and Development of Wild-Type and Starch-Deficient Mutants of Arabidopsis During Spaceflight," *Physiologia Plantarum* 102, no. 4 (1998): 493–502.

104 *Peter Boysen-Jensen expanded on*: Peter Boysen-Jensen, "Über die Leitung des phototropischen Reizes in der Avenakoleoptile," *Berichte des Deutschen Botanischen Gesellschaft* 31 (1913): 559–66.

107 *As the Polish scientist Maria Stolarz*: Maria Stolarz et al., "Disturbances of Stem Circumnutations Evoked by Wound-Induced Variation Potentials in *Helianthus annuus* L.," *Cellular & Molecular Biology Letters* 8, no. 1 (2003): 31–40.

108 *This hypothesis remained unchallenged*: Anders Johnsson and Donald Israelsson, "Application of a Theory for Circumnutations to Geotropic Movements," *Physiologia Plantarum* 21, no. 2 (1968): 282–91.

109 *Brown had to wait*: Allan H. Brown et al., "Circumnutations of Sunflower Hypocotyls in Satellite Orbit," *Plant Physiology* 94 (1990): 233–38.

109 *Sunflower seedlings exhibit robust*: John Z. Kiss, "Up, Down, and All Around: How Plants Sense and Respond to Environmental Stimuli," *Proceedings of the National Academy of Sciences of the United States of America* 103, no. 4 (2006): 829–30.

110 *A few years ago*: Kitazawa et al., "Shoot Circumnutation and Winding Movements Require Gravisensing Cells."

111 *Anders Johnsson could put*: Anders Johnsson, Bjarte Gees Solheim, and Tor-Henning Iversen, "Gravity Amplifies and Microgravity Decreases Circumnutations in *Arabidopsis thaliana* Stems: Results from a Space Experiment," *New Phytologist* 182, no. 3 (2009): 621–29.

111 *The overall mechanism of*: Morita, "Directional Gravity Sensing in Gravitropism."

SIX. WHAT A PLANT REMEMBERS

114 *Mark Jaffe, the same scientist*: Mark J. Jaffe, "Experimental Separation of Sensory and Motor Functions in Pea Tendrils," *Science* 195, no. 4274 (1977): 191–92.

115 *Tulving proposed that human memory*: Endel Tulving, "How Many Memory Systems Are There?," *American Psychologist* 40, no. 4 (1985): 385–98. While Tulving's models of memory are well recognized, they should not be accepted as monolithic, and within the field of memory numerous models and theories exist, not all of which are mutually exclusive.

115 *But plants are capable of sensing*: Fatima Cvrckova, Helena Lipavska, and Viktor Zarsky, "Plant Intelligence: Why, Why Not, or Where?," *Plant Signaling & Behavior* 4, no. 5 (2009): 394–99.

116 *What's fascinating is that the latest*: Todd C. Sacktor, "How Does PKMz Maintain Long-Term Memory?," *Nature Reviews Neuroscience* 12, no. 1 (2011): 9–15.

117 *Scientists have been puzzled*: John S. Burdon-Sanderson, "On the Electromotive Properties of the Leaf of *Dionaea* in the Excited and Unexcited States," *Philosophical Transactions of the Royal Society of London* 173 (1882): 1–55.

117 *A century later, Dieter Hodick and Andreas Sievers*: Dieter Hodick and Andreas Sievers, "The Action Potential of *Dionaea muscipula* Ellis," *Planta* 174, no. 1 (1988): 8–18.

118 *Alexander Volkov and his colleagues*: Alexander G. Volkov, Tejumade Adesina, and Emil Jovanov, "Closing of Venus Flytrap by Electrical Stim-

ulation of Motor Cells," *Plant Signaling & Behavior* 2, no. 3 (2007): 139–45.

119 *When Volkov pretreated his plants*: Ibid.

119 *In the mid-twentieth century*: Rudolf Dostál, *On Integration in Plants*, translated by Jana Moravkova Kiely (Cambridge, Mass.: Harvard University Press, 1967).

122 *But Dostál noticed that if he*: Described in Anthony Trewavas, "Aspects of Plant Intelligence," *Annals of Botany* 92, no. 1 (2003): 1–20.

122 *Thellier, a member of the French*: Michel Thellier et al., "Long-Distance Transport, Storage, and Recall of Morphogenetic Information in Plants: The Existence of a Sort of Primitive Plant 'Memory,'" *Comptes Rendus de l'Académie des Sciences, Série III* 323, no. 1 (2000): 81–91.

124 *Trofim Denisovich Lysenko was notorious*: E. W. Caspari and R. E. Marshak, "The Rise and Fall of Lysenko," *Science* 149, no. 3681 (1965): 275–78.

125 *Other scientists also knew*: John H. Klippart, *Ohio State Board of Agriculture Annual Report* 12 (1857): 562–816.

127 *These investigations have highlighted*: Ruth Bastow et al., "Vernalization Requires Epigenetic Silencing of FLC by Histone Methylation," *Nature* 427, no. 6970 (2004): 164–67; Yuehui He, Mark R. Doyle, and Richard M. Amasino, "PAF1-Complex-Mediated Histone Methylation of *Flowering Locus C* Chromatin Is Required for the Vernalization-Responsive, Winter-Annual Habit in *Arabidopsis*," *Genes & Development* 18, no. 22 (2004): 2774–84.

128 *How this occurs, and how it's regulated*: Pedro Crevillen and Caroline Dean, "Regulation of the Floral Repressor Gene *FLC*: The Complexity of Transcription in a Chromatin Context," *Current Opinion in Plant Biology* 14, no. 1 (2011): 38–44.

129 *Barbara Hohn's laboratory in Basel*: Jean Molinier et al., "Transgeneration Memory of Stress in Plants," *Nature* 442, no. 7106 (2006): 1046–49.

130 *Igor Kovalchuk created a follow-up*: Alex Boyko et al., "Transgenerational Adaptation of *Arabidopsis* to Stress Requires DNA Methylation and the Function of Dicer-Like Proteins," *PLoS One* 5, no. 3 (2010): e9514.

131 *Hohn's results were not*: Ales Pecinka et al., "Transgenerational Stress Memory Is Not a General Response in Arabidopsis," *PLoS One* 4, no. 4 (2009): e5202.

131 *The growing consensus, however*: Eva Jablonka and Gal Raz, "Transgenerational Epigenetic Inheritance: Prevalence, Mechanisms, and Implications for the Study of Heredity and Evolution," *Quarterly Review of Biology* 84, no. 2 (2009): 131–76; Faculty of 1000, evaluations, dissents,

and comments for Molinier et al., "Transgeneration Memory of Stress in Plants," Faculty of 1000, September 19, 2006, F1000.com/1033756; Ki-Hyeon Seong et al., "Inheritance of Stress-Induced, ATF-2-Dependent Epigenetic Change," *Cell* 145, no. 7 (2011): 1049–61.

131 *In all cases, this "memory"*: Tia Ghose, "How Stress Is Inherited," *Scientist* (2011), http://the-scientist.com/2011/07/01/how-stress-is-inherited.

132 *It was a great surprise*: Eric D. Brenner et al., "Arabidopsis Mutants Resistant to S(+)-Beta-Methyl-Alpha, Beta-Diaminopropionic Acid, a Cycad-Derived Glutamate Receptor Agonist," *Plant Physiology* 124, no. 4 (2000): 1615–24; Hon-Ming Lam et al., "Glutamate-Receptor Genes in Plants," *Nature* 396, no. 6707 (1998): 125–26.

132 *At this point we still*: Erwan Michard et al., "Glutamate Receptor–Like Genes Form Ca2+ Channels in Pollen Tubes and Are Regulated by Pistil D-Serine," *Science* 332, no. 434 (2011).

132 *Tulving further proposed*: Tulving, "How Many Memory Systems Are There?"

133 *"lowest level of"*: Cvrckova, Lipavska, and Zarsky, "Plant Intelligence."

EPILOGUE: THE AWARE PLANT

135 *Everyone from Alfred Binet*: Alfred Binet, Théodore Simon, and Clara Harrison Town, *A Method of Measuring the Development of the Intelligence of Young Children* (Lincoln, Ill.: Courier, 1912); Howard Gardner, *Intelligence Reframed: Multiple Intelligences for the 21st Century* (New York: Basic Books, 1999); Stephen Greenspan and Harvey N. Switzky, "Intelligence Involves Risk-Awareness and Intellectual Disability Involves Risk-Unawareness: Implications of a Theory of Common Sense," *Journal of Intellectual and Developmental Disability*, in press (2011); Robert J. Sternberg, *The Triarchic Mind: A New Theory of Human Intelligence* (New York: Viking, 1988).

135 *While some researchers consider*: Reuven Feuerstein, "The Theory of Structural Modifiability," in *Learning and Thinking Styles: Classroom Interaction*, edited by Barbara Z. Presseisen (Washington, D.C.: NEA Professional Library, National Education Association, 1990); Reuven Feuerstein, Refael S. Feuerstein, and Louis H. Falik, *Beyond Smarter: Mediated Learning and the Brain's Capacity for Change* (New York: Teachers College Press, 2010); Binyamin Hochner, "Octopuses," *Current Biology* 18, no. 19 (2008): R897–98; Britt Anderson, "The G Factor in Nonhuman Animals," *Novartis Foundation Symposium* 233 (2000): 79–90, discussion 90–95.

135 *"It appears to me"*: William Lauder Lindsay, "Mind in Plants," *British Journal of Psychiatry* 21 (1876): 513–32.

135 *Anthony Trewavas, an esteemed*: Anthony Trewavas, "Aspects of Plant Intelligence," *Annals of Botany* 92, no. 1 (2003): 1–20.

136 *Controversy arose among plant biologists*: Eric D. Brenner et al., "Plant Neurobiology: An Integrated View of Plant Signaling," *Trends in Plant Science* 11, no. 8 (2006): 413–19.

136 *Some of these similarities*: Frantiöek Baluöka, Simcha Lev-Yadun, and Stefano Mancuso, "Swarm Intelligence in Plant Roots," *Trends in Ecology and Evolution* 25, no. 12 (2010): 682–83; Frantiöek Baluöka et al., "The 'Root-Brain' Hypothesis of Charles and Francis Darwin: Revival After More Than 125 Years," *Plant Signaling & Behavior* 4, no. 12 (2009): 1121–27; Elisa Masi et al., "Spatiotemporal Dynamics of the Electrical Network Activity in the Root Apex," *Proceedings of the National Academy of Sciences of the United States of America* 106, no. 10 (2009): 4048–53.

136 *Many other biologists who*: Amedeo Alpi et al., "Plant Neurobiology: No Brain, No Gain?," *Trends in Plant Science* 12, no. 4 (2007): 135–36.

139 *The International Association for the Study of Pain*: John J. Bonica, "Need of a Taxonomy," *Pain* 6, no. 3 (1979): 247–52. See also www.iasp-pain.org/AM/Template.cfm?Section=Pain_Definitions&Template=/CM/HTMLDisplay.cfm&ContentID=1728#Pain.

139 *Indeed, even in humans*: Michael C. Lee and Irene Tracey, "Unravelling the Mystery of Pain, Suffering, and Relief with Brain Imaging," *Current Pain and Headache Reports* 14, no. 2 (2010): 124–31.

140 *For example, in 2008*: Alison Abbott, "Swiss 'Dignity' Law Is Threat to Plant Biology," *Nature* 452, no. 7190 (2008): 919.

Acknowledgments

What a Plant Knows would never have been published without the input of three amazing women.

First, my wife, Shira, who encouraged me to push the envelope, do something beyond academic research and writing, and, finally, to press "send." Without her love and belief, this book would never have happened.

Second, my agent, Laurie Abkemeier. Her experience, tenacity, support, and boundless optimism made a naive author feel like a Pulitzer Prize–winning veteran. I was fortunate in finding not only an agent but a friend.

Third, my editor at Scientific American / Farrar, Straus and Giroux, Amanda Moon, who had the daunting task of turning my academic wording into readable prose. Amanda worked tirelessly to edit and reedit each chapter, and then do it again a third, a fourth, and a fifth time, all with utmost patience.

Many scientists from around the world helped me craft this into a scientifically valid work. Professors Ian Baldwin (Max Planck Institute for Chemical Ecology), Janet Braam (Rice University), John Kiss (Miami University), Viktor Zarsky (Academy of Sciences of the Czech Republic), and Eric Brenner (New York University) were kind enough to take time from their busy schedules to read parts of this book and to make sure that the science was fairly presented. The idea for this book originated in discussions with Eric, and I will be forever thankful for his insight, encouragement, and friendship. I also thank Professor Jonathan Gressel (Weizmann Institute of Science), Dr. Lilach Hadany (Tel Aviv University), Professor Anders Johnsson (Norwegian University of Science and Technology), Professor Igor Kovalchuk (University of Lethbridge), and Dr. Virginia Shepherd (the University of New South Wales) for input at various stages of this project. The influence of my mentors, Professor Joseph Hirschberg and Professor Xing-Wang Deng, is felt in all the science I do and write.

I thank Karen Maine for her edits and diligence in keeping me on time, Ingrid Sterner for an amazing copyedit, and the team at Scientific American / Farrar, Straus and Giroux, who were wonderful to work with.

I am lucky to have fabulous colleagues at Tel Aviv University who provided many helpful hallway discussions and ideas. In particular, many of the ideas in this book were explored first with Professors Nir Ohad and Shaul Yalovsky in our course Introduction to Plant Sciences. I want to thank my labmates, Ofra, Ruti, Sophie, Elah, Mor, and Giri, for accepting my absences in overseeing their research while I wrote this book and especially Dr. Tally Yahalom, who covered for me in running the lab. My daily interaction with them is a constant reminder of why research is so exciting. I'm also indebted to the benefactor of the Manna Center for Plant Biosciences who has helped show me how modesty coupled with focus can synergize to reach important goals.

I wish to thank Alan Chapelski for the portrait, and Deborah Luskin, who helped me start to write. My immediate and extended families have been a source of unending support. From my sister, Raina, to Ehud, Gitama, Yanai, Phyllis, and my mother,

Marcia, who were the first readers of the manuscript, I am forever thankful. My children, Eytan, Noam, and Shani, are constant sources of joy and were even available to provide me with a missing word. And lastly, my father, David, who offered me edits and constant support, and lived vicariously through the book's publication.

Index

Page numbers in *italics* refer to illustrations.

ILLUSTRATION CREDITS

13 Amédée Masclef, *Atlas des plantes de France* (Paris: Klincksieck, 1891).

14 Varda Wexler.

16 Ernst Gilg and Karl Schumann, *Das Pflanzenreich*, Hausschatz des Wissens (Neudamm: Neumann, ca. 1900).

22 USDA-NRCS PLANTS Database / Nathaniel Lord Britton and Addison Brown, *An Illustrated Flora of the Northern United States, Canada, and the British Possessions*, 3 vols. (New York: Charles Scribner's Sons, 1913), 2:176.

32 USDA-NRCS PLANTS Database / Nathaniel Lord Britton and Addison Brown, *An Illustrated Flora of the Northern United States, Canada, and the British Possessions*, 3 vols. (New York: Charles Scribner's Sons, 1913), 3:49.

36 Prof. Dr. Otto Wilhelm Thomé, *Flora von Deutschland, Österreich, und der Schweiz* (Gera: Köhler, 1885).

37 Walter Hood Fitch, *Illustrations of the British Flora* (London: Reeve, 1924).

40 Francisco Manuel Blanco, *Flora de Filipinas* [Atlas II] (Manila: Plana, 1880–83).

42 Modified from figures 2 and 3, in Martin Heil and Juan Carlos Silva Bueno, "Within-Plant Signaling by Volatiles Leads to Induction and Priming of an Indirect Plant Defense in Nature," *Proceedings of the National Academy of Sciences of the United States of America* 104, no. 13 (2007): 5467–72. Copyright © 2007 National Academy of Sciences, U.S.A.

46 Illustration from a photograph of *Amorphophallus titanum* in Wilhelma by Lothar Grünz (2005).

51 USDA-NRCS PLANTS Database, http://plants.usda.gov, accessed August 25, 2011, National Plant Data Team, Greensboro, N.C., 27401-4901 U.S.A.

54 Taken from figure 12, in Charles Darwin, *Insectivorous Plants* (London: John Murray, 1875).

59 Paul Hermann Wilhelm Taubert, *Natürliche Pflanzenfamilien* (Leipzig: Engelmann, 1891), 3:3.

61 USDA-NRCS PLANTS Database / Nathaniel Lord Britton and Addison Brown, *An Illustrated Flora of the Northern United States, Canada, and the British Possessions*, 3 vols. (New York: Charles Scribner's Sons, 1913), 3:345.

67 USDA-NRCS PLANTS Database / Nathaniel Lord Britton and Addison Brown, *An Illustrated Flora of the Northern United States, Canada, and the British Possessions*, 3 vols. (New York: Charles Scribner's Sons, 1913), 3:168.

76 George Crouter, in Dorothy L. Retallack, *The Sound of Music and Plants* (Santa Monica, Calif.: DeVorss, 1973), p. 6.

79 Francisco Manuel Blanco, *Flora de Filipinas*, book 4 (Manila: Plana, 1880–83).

81 Prof. Dr. Otto Wilhelm Thomé, *Flora von Deutschland, Österreich, und der Schweiz* (Gera: Köhler, 1885).

96 Varda Wexler.

98 Taken from figure 196, in Charles Darwin and Francis Darwin, *The Power of Movement in Plants* (New York: D. Appleton, 1881).

101 Walter Hood Fitch, *Curtis's Botanical Magazine* vol. 94, ser. 3, no. 24 (1868), plate 5720.

105 USDA-NRCS PLANTS Database / A. S. Hitchcock, revised by Agnes Chase, *Manual of the Grasses of the United States*, USDA Miscellaneous Publication no. 200 (Washington, D.C.: U.S. Government Printing Office, 1950).

107 Taken from figure 6, in Charles Darwin and Francis Darwin, *The Power of Movement in Plants* (New York: D. Appleton, 1881).

108 USDA-NRCS PLANTS Database / USDA Natural Resources Conservation Service, *Wetland Flora: Field Office Illustrated Guide to Plant Species*.

120 Varda Wexler.

121 USDA-NRCS PLANTS Database / Nathaniel Lord Britton and Addison Brown, *An Illustrated Flora of the Northern United States, Canada, and the British Possessions*, 3 vols. (New York: Charles Scribner's Sons, 1913), 2:436.

123 USDA-NRCS PLANTS Database / Nathaniel Lord Britton and Addison Brown, *An Illustrated Flora of the Northern United States, Canada, and the British Possessions*, 3 vols. (New York: Charles Scribner's Sons, 1913), 3:497.

braiding
Sweetgrass